Growing Tropical Plants

Growing Tropical Plants

John Mason

Kangaroo Press

Acknowledgments

Editorial and Research Assistants
The following staff from the Australian Correspondence Schools have contributed to the preparation of this book:
Iain Harrison
Paul Plant
Kathy Travis
Andrew Penney
Rosemary Lawrence

Thanks also go to the following people for their advice and information on selected plants:
George Brown, former Parks Manager and Mayor of Darwin.
Roger Spencer, National Herbarium, South Yarra, Victoria.

Reprinted 2001
First published in 1997 by Kangaroo Press Pty Ltd
an imprint of Simon and Schuster (Australia)
20 Barcoo Street (PO Box 507)
East Roseville, 2069, Australia
Printed in Hong Kong through Colorcraft Ltd

ISBN 0 86417 8093

Contents

Introduction

This book describes how to grow plants commonly cultivated in the tropics and the subtropics. Many of the plants listed here can also be grown successfully in mild-temperate climates, as indoor plants, or in protected warm parts of a garden.

GETTING TO KNOW PLANTS

The bulk of the book deals with different kinds (genera) of cultivated plants suitable to at least some parts of the tropics. The information for each of these groups is arranged under four headings:

Appearance: The characteristics that are usually representative of the genus.

Establishment: How and where to start the plant growing in the garden.

Culture: The most important things you need to know about keeping the plant growing and healthy.

Cultivars: Some of the more important differences from the genus between selected species, hybrids and/or varieties which are cultivated in tropical and subtropical areas.

WHAT DO WE MEAN BY TROPICAL AND SUBTROPICAL?

For the purposes of this book, a **tropical** plant is one that will grow successfully between the tropic of Cancer and the Tropic of Capricorn. Tropical climates are found in South-east Asia, much of India, northern Australia, Central America, the Carribbean, the northern parts of South America, many of the Pacific islands and perhaps the central half of the African continent. Tropical areas in general have the highest average temperature levels, the longest frost-free growing seasons and the greatest amount of light (intensity and duration) compared with other regions on the planet.

Subtropical generally refers to the regions between about 23° and 30° south and north of the equator. These regions generally have lower average temperatures, shorter frost-free growing seasons and less light overall than tropical regions, but without the cold winters of the temperate zone. Subtropical climates are found in Australia (eg. south-east Queensland), Africa (eg. northern South Africa), the USA (eg. parts of Florida, Louisiana, southern California, Texas), etc.

In writing this book, I have tried to give you value for money (covering more plant varieties than other books of a similar size) and general guidelines which should enable you to avoid most of the major problems that confront gardeners in the tropics and sub-tropics. I wish you and your garden great success.

J.M.
July, 1997

1

Tropical Gardens

Tropical gardens can vary from dry and desert-like, to coastal, to dense, lush and leafy environments resembling the dynamic workings of a rainforest. Many different garden effects are possible using tropical plants. You can also, on a small scale, transform a garden into a microclimate using tropical plants that are not normally found in your locality.

TYPES OF CLIMATES

It is important to understand that there are variations in climate within tropical and subtropical regions, and that not all tropical plants like the same conditions. These are just some of the different climates which 'tropical plants' come from:

Seasonal or Constant

Many tropical climates have two distinct seasons: a wet season (monsoon) with heavy rains and high humidity, and a dry season, when there is little rain. Some parts of the tropics are much less seasonal, with temperature and rainfall conditions remaining similar all year round (eg. Honolulu and some other parts of Hawaii).

Mountain

Mountains can be high and cold in some parts of the tropics. There are even snow-covered mountains in tropical New Guinea. Plants which grow on tropical mountains can be quite different to plants that grow in tropical lowlands.

Savannas (grasslands)

Savannas are tropical or subtropical open grasslands that are bordered either by rainforest or swamps, or a mixture of both. They tend to experience a greater variation in temperature than coastal or rainforest environments. They can be very humid and damp environments due to high moisture and open sunny conditions. Broadleaved ground covers with bright flowers are very much at home in this environment.

Arid

Some areas that are tropical have wide extremes when it comes to moisture availability. Arid zones usually have poor soils such as claypans, sands or gravels, and are subject to occasional downpours that can result in flooding. Therefore, the plants that thrive in this type of environment are either very hardy or have very specialised survival strategies to meet such conditions. Arid environments can be much colder at night and hotter during the day than other parts of the tropics or sub-tropics.

Rainforest

The rainforest is one of the most dynamic environments that exist, with an abundance of species that often cannot be found elsewhere. Conditions throughout a rainforest are constantly changing as the forest undergoes a competitive ageing process that sees many species unable to survive. Thus those that succeed do so as a result of highly specialised ecological refinements.

Coastal

Coastal regions can be a mixture of all the previously mentioned environments with the added complications of salt-laden spray, periods of heavy winds, nutrient-poor soils and, at times, encroaching sand dunes. Plants that grow in coastal areas need to be quite hardy to survive such conditions. Being close to water has a buffering effect upon temperature extremes—coastal areas do not get as hot or as cold as environments in similar latitudes that are further away from water.

The good news and the bad news

Warm climates can create both good and bad conditions for gardeners in a tropical climate, where—

- Plants tend to grow faster, taller, more lushly and last longer.
- Pests and diseases also tend to grow bigger and faster, but if plants are relatively healthy they are able to recover faster from these problems in a warm climate.
- Humidity can be higher and more prolonged than in cool climates.
- Winds and storms can be more forceful.
- Soil can dry out faster.
- Foliage can suffer sunburn more readily.

Generally speaking, this means that—

- Often, plants need to be grown differently in the tropics than they are in the sub-tropics, or in a temperate climate as an indoor plant.
- In most cases the same plant grows bigger in the tropics than in the sub-tropics, and bigger in the sub-tropics than in temperate climates.
- Drainage is very important in tropical areas to avoid roots being flooded in heavy rainfalls.
- Plants susceptible to damage from water-logging might be better planted on mounds or slopes.
- Plants need to be inspected for pests and diseases more often and action taken immediately. In a cool climate you might wait for a few days or weeks to see whether insects develop into a serious problem, but in the tropics they can develop from a minor to a serious problem overnight.

SOILS AND FERTILISERS

Different plants like different types of soils. Some plants will only grow well in acidic soils, others only in alkaline soils. You can tell whether a soil is acidic or alkaline by testing it with a pH test kit or probe, available from many nurseries and garden centres.

Organic material such as mulch or manure will make a soil more acidic (rainforests often have acidic soils). Lime, shell grit or coral will make a soil more alkaline (alkaline soils are often found on coral islands or in beach areas containing lots of shell grit).

Where plants mentioned in the plant descriptions of this book have specific requirements for acidic or alkaline conditions, it will be mentioned under cultural details for that plant variety.

Soil problems

- Many tropical plants need good drainage. Without it, the roots tend to rot very rapidly. In general they prefer well drained soils on slopes, sandy soil, or well structured (ie. friable) soils.
- Heavy rains in tropical areas will leach nutrients from the soil faster than in most temperate climates.
- High temperatures and humidity often cause fertilisers to 'dump' their nutrients within a very short time. Accordingly, so-called 'slow release' fertilisers may actually release their nutrients very quickly.

Improving soils

Before deciding how to, or even whether to improve a soil, you need to know if it is good, bad or in need of improvement.

Drainage can be tested easily by observing the way in which water moves through soil. Place a sample of the soil in a pot and then water it, watching how quickly or slowly the water passes through. Keep in mind that this is a simple test. When soil is disturbed by digging, its characteristics may change. However, this test will give you a general picture of your soil's drainage capacity. Periodic sampling of the soil will enable you to understand its drainage characteristics.

Soil nutrition is, to some extent, indicated by the vigour of plants growing in a particular soil.

Nutrition can easily be improved by the addition of fertilisers, and, over the long term, by the addition of organic matter.

A well structured soil usually has a crumbly nature, with plenty of pore space (voids) between the small crumbs. These types of soils are generally easily cultivated, have good drainage, and good aeration (oxygen is needed by plant roots). Anything that destroys this crumbly structure, such as over-cultivation, regular trampling, or traffic of vehicles should be avoided.

Ways to improve soils include:

- Adding sand to clay soils to improve drainage. This is only practicable on a small scale (eg. garden beds) as a lot of sand would be required to have a reasonable effect, and it would also need to be well mixed.
- Adding clay to sandy soil to improve its ability to hold water. Much less clay is required to improve the water-holding ability of sand than when adding sand to help improve drainage in clay. Thorough mixing is also important in this case.
- Adding organic matter to any soil, except those rare ones already high in organic matter, will help improve soil structure and hence drainage. It will also improve soil fertility, soil moisture-holding capacity, and will provide a buffer against sudden chemical or temperature changes in the soil.
- Adding soil ameliorants—lime can be added to help improve structure in clay soils and/or to raise pH levels in acidic soils; gypsum can be added to help improve structure in clay soils, without affecting pH significantly, or to help improve saline-sodic soils by displacing sodium ions from soil particles so that they can be leached out of the soil.
- Using acidifying fertilisers (eg. ammonium sulphate, ammonium phosphate) will help lower soil pH, as well as provide valuable nutrients. Repeated applications of organic matter may also have an acidifying effect over time. On a small scale, sulphur dust can be used to lower soil pH, but this is quite expensive.
- Mulching will help protect the soil from erosion and compaction, control weeds, and help protect plant roots from temperature extremes. Organic mulches will add valuable organic matter to the soil as they decompose.

WATER MANAGEMENT

Generally, water plants well when they are growing and reduce water when growth slows or becomes dormant. Excess water may result in the roots rotting (a particular problem with true tropicals when grown in cooler climates, even under cover).

The heat of a tropical or subtropical climate will cause water to evaporate quickly from the surface of the soil, so frequent light waterings may achieve very little. Irrigation methods must be designed and used to wet the deeper layers of the soil. This can be done by:

1. Watering slowly for a long period, so that the water soaks in (eg. a drip irrigation system).
2. Drilling holes beside plants (eg. with a narrow auger), being careful to minimise any root damage, inserting a pipe into the hole to stop it collapsing, and then watering into the pipe. In this way the water begins to enter the ground from deep down, rather than near the surface.

Additions of soil ameliorants such as liquid soil-wetters aid in water conservation. These products allow soils to hold additional water so that less frequent irrigations are required.

Where high levels of humidity are required to maintain healthy plant foliage, apply frequent light applications of water directly to the foliage by such methods as overhead sprinklers or misting systems (using nozzles or sprinklers that give small droplet sizes).

LIGHT

Light requirements will vary considerably from one tropical plant species to another. Those that grow naturally in open sunny conditions, such as in savanna or in the upper canopy of rainforests, generally require plenty of light. By comparison, the large percentage of plants that live beneath the protective upper canopy of rainforests generally do best in filtered sunlight or partial shade. Many of these plants have large broad leaves to maximise the amount of light they receive, and are very decorative. It is from this group that many of the plants grown indoors in cooler climates are derived.

When deciding how much light your plants require, the best indication is to find out their normal growing conditions in the wild. If you find, when growing certain plants, that they are becoming tall and spindly, it is likely they require more light. If you find they are getting a bit sunburnt, or their leaf tips are perhaps showing signs of scorching, then you may need to provide some sun protection.

PLANTING

The first step in planting a garden is to choose the right plants. This can be a difficult task, even with expert advice. There are so many choices, and these often come down to personal preference. To avoid problems, consider the following:

- Avoid placing plants that have invasive roots near buildings, paved areas, water features, drains, and septic tanks (e.g. eucalypts, ficus).
- Don't use plants that have the potential to become weeds. This includes plants that seed prolifically, are prone to sucker, or are rampant creepers (e.g. wandering jew, lantana, even common jasmine).
- Consider how big each plant is likely to grow. You should allow sufficient space for future growth. Avoid planting large trees in places where they will eventually create problems (e.g. shade out lawns or other plants, damage buildings).
- Don't use trees that are likely to drop branches (e.g. many eucalypts).
- Choose plants that are likely to withstand local storms, pests and diseases, or poor soils.
- Select plants suited to the tropics or sub-tropics. Many people become disheartened when a beloved plant dies, when all to often it was a plant that had little chance of surviving in the climate.

A SIMPLE APPROACH

The best way to approach planting is often the simplest. Consider the following strategies:

- Keep plants together in compatible groups. Keep foliage plants together, herbs together, orchids together or, alternatively, group together those plants that require similar amounts of watering, or similar amounts of shade.
- Let the fittest survive.
- Mass plant areas (ie. several of the one variety together) to form hedges or eye-catching displays.
- Be prepared to get rid of plants that don't work (either don't look good or don't grow well). There's no harm in being ruthless.
- Use reliable plants that neighbours are growing successfully, unless you wish to experiment. If so, research the requirements of your plants well, or be prepared to lose a few along the way.

PROPAGATING PLANTS

There are two broad groups of propagating techniques used in the production of plants:

Sexual propagation involves growing a plant from a seed or spore which has been produced by fertilisation of the female part of one plant by the male part of either the same plant or, more commonly, of another plant. Plants grown this way can have some characteristics of one parent and different characteristics of the other parent. Sexually propagated plants are never exactly the same as the plant from which the seed or spores were taken.

Asexual, or **vegetative** propagation involves reproducing a new plant from only one parent. A part of an existing plant is treated in such a way that it can produce a new plant, i.e. asexual propagation involves growing a new plant from a piece of stem, leaf or root. Plants produced by asexual means are genetically identical to the parent plant. This is sometimes described as 'cloning'. Common vegetative methods of propagation include the use of cuttings, budding and grafting, division and separation. As cutting propagation is the most common vegetative method, we will explore this technique in greater detail.

CUTTINGS

A cutting is a piece of vegetative growth (non-sexual, ie. not the flower or fruit) which is detached from a plant and treated in such a way as to stimulate it to grow roots, stems and leaves, thus producing a new plant. Cutting propagation can be carried out on a very wide variety of plants.

The importance of cuttings

Plants are reproduced by cuttings for a number of reasons:

- A cutting-grown plant is genetically identical to the parent plant (ie. the plant from which the original cutting was taken). This is not necessarily so when plants are grown from seed. Propagation by cuttings is the most widely used technique for reproducing 'true to type' plants, ensuring that the unique characteristics of the parent plant are reflected in the progeny.
- Cuttings can often by used to propagate plants that:

a. do not produce viable seed, or produce seed at irregular times;
b. have seed that is difficult to germinate (eg. Boronia, Eriostemon);
c. have seed that is difficult to collect—for example, plants that have seed pods that burst open, dispersing the seeds widely;
d. produce their seed at a time when seed cannot be collected, or collection would require a further trip to the area (often very difficult for remote areas), or can only be collected with difficulty (eg. plants whose seed matures during wet seasons, when access may be limited).

Types of cuttings

Cuttings can be classified in two different ways:

- According to the time of year the cutting is taken (or the stage of growth of the plant when it is taken). For example, a softwood cutting is one taken in spring when the young growth on the plant is soft tissue.
- According to the part of the plant that is used. For example, a leaf cutting is a cutting made from just a leaf, or part of a leaf).

Softwood cuttings

These are stem cuttings taken from new growth that is soft. This commonly occurs during spring, but may occur at other times of the year if suitable material is available.

Hardwood cuttings

Stem cuttings taken in winter from old growth that is hard. This type of cutting is more commonly used on deciduous plants from cooler regions.

Semi-hardwood cuttings

Stem cuttings, usually taken in late summer or early autumn when recent spring growth is in the process of hardening. Many shrubs are propagated this way.

Herbaceous cuttings

Leafy stem cuttings taken from a soft-wooded (succulent growing) plant. These can be taken virtually at any time of the year.

Stem cuttings

These are derived from a section of stem, usually (but not always) with some leaves left on the top but with lower leaves removed. There should be a node (this is a point at which a bud emerges) at the bottom of the cutting and another node at the top of the cutting. There may be one node, or several, in between.

Tip cuttings

Stem cuttings taken from the growing tip of a plant. Softwood cuttings are often tip cuttings.

Heel cuttings

A heel cutting is a stem cutting of one-year-old wood which has, attached to the base, a small section of two-year-old wood. This section of older wood is called a heel. Heel cuttings are normally obtained by tearing side shoots from a small branch or stem; the torn section is then trimmed neatly with secateurs or a knife.

Nodal cuttings

A stem cutting without a heel, where the base of the cutting is made as a right-angle cut just below a node.

Basal cuttings

A stem cutting where the base of the cutting is made at the point where the young shoot joins the older branch. At this point there is often some swelling in the stem. Unlike the heel cutting, the basal cutting does not necessarily contain any older wood.

Leaf bud cuttings

A full leaf (leaf blade and stalk) with a small piece of the stem to which the leaf was attached. At the junction of the stem there is a bud (which is retained).

Leaf cuttings

Either a section of a leaf or a full leaf, including the leaf stalk (petiole). In the case of a section of a leaf being used (eg. African violet, Peperomia), the cutting must include part of a major leaf vein. New growth (a shoot and root) will normally grow from the base of the cutting (ie. the base of the leaf stalk, or the base of the leaf vein). African violets are commonly grown from full leaves.

Cane cuttings

A small section of cane from the plant, containing only one or two nodes and no leaves, is inserted horizontally (instead of vertically, as cuttings are normally done), with a bud showing just above the surface of the medium. This is used with plants such as Dracaena and Diffenbachia, where it is difficult to obtain large quantities of cutting material. Heating and misting are usually essential for commercial success.

Root cuttings

Sections of relatively young (1–3 years old) root, 2–10cm long, taken preferably from young plants. Cuttings are planted horizontally 2–4cm deep in the propagation medium. Small, delicate roots should be shorter and planted shallower (maybe with just 1cm layer of sand over the top). Larger roots can be longer and planted deeper.

Cutting length

For stem cuttings (including softwood, hardwood, etc.), the cutting length will vary according to the number of buds on the stem. Cuttings from plants that have short internode lengths (short distances between consecutive buds) are often as short as 5–7cm long, while cuttings that have longer distances between nodes (where buds are found) will often be up to 15cm or more long. The important thing is to be sure to have at least 4–6 buds present on each cutting.

Taking cuttings

When taking cuttings, consider the following:

- Cuttings should be taken from healthy parent material.
- They should be cut with a sharp knife, secateurs or similar tool to make a clean, neat cut. These tools should be dipped in disinfectant regularly to prevent the spread of pests and diseases.
- Avoid taking cuttings during the warm part of the day.
- Prepare your cuttings immediately, or store them in a position where they are kept cool and moist until you are ready (ideally within a few hours).
- The basal cut on a cutting should be made no more than about 6 or 7mm below a node.
- If the tip of a stem is not being used as the cutting, then the top cut should be made just above a node (this is to prevent die-back).
- Generally, the lower two-thirds to three-quarters of the leaves should be removed, taking care not to rip any bark. For hardwood cuttings, all leaves are usually removed. Any flowers or flower buds should also be removed.
- The cuttings should be placed into a suitable container, such as a 15cm diameter plant pot that contains a suitable propagating mix. A

suitable mix for most plants is 50% coarse washed sand and 50% sieved peat mix. This will provide good drainage while ensuring sufficient moisture is retained to keep the plant healthy until roots have formed. Cuttings are usually placed into the mix by making a hole with a narrow stick, lowering the cutting into the hole to about one-half to two-thirds its length, and firming the mix around the cutting.

- The cuttings should then be watered well and placed in a warm, protected position in indirect light.
- To improve strike rates, the base of the cuttings can be dipped into a hormone compound before being placed into the propagation medium.
- The container of cuttings can be placed into a structure such as a greenhouse or cold frame to provide improved growing conditions (e.g. higher temperatures and humidity) or, alternatively, a clear plastic bag can be placed over each pot to create a mini-glasshouse effect.
- Be sure to keep the cuttings sufficiently watered, but not overwatered.
- Remove the cuttings once roots start appearing through the base of the container, and pot each cutting up into an individual container.

SEED

Many plants can be easily propagated by seed providing that the seed you have collected is at the right stage of maturity (ripe, but not over-ripe) and that you provide a suitable environment for germination. The same propagation mix used for striking cuttings can be used to germinate most seeds (50% coarse washed sand and 50% sieved peat moss). For plants that might like conditions a bit moister, add more peat moss and vice versa. Broad plant pots or seedling trays are suitable containers. The propagating mix should be placed into the container and lightly firmed down, leaving enough distance below the top of the container to allow for water retention during watering.

The seeds can be sown on the surface of the mix and lightly covered. Some types don't need covering at all. Water thoroughly, but lightly, to ensure that the seed isn't dislodged, and place the container in a warm, protected position in indirect light. As with cuttings, a glasshouse or cold frame or similar structure can help provide improved conditions. Seedlings can be carefully removed once they have their second or third pair of leaves and potted up into a suitable potting mix or soil.

Some seeds, particularly from rainforest areas, may require some sort of pre-sowing treatment before they will germinate. This could include allowing the seed to ferment in its surrounding fruit, or by abrasion of hard seed coats to thin them down.

HOT BEDS AND MISTING

Propagating beds that provide basal heat to propagation containers (both seed and cuttings) can help speed up the rate of root formation or seed germination. Many such beds use hot water pipes or electric heating cables to provide the heat, and the temperature can often be controlled thermostatically to give a preferred temperature range.

Misting systems can also be used to help reduce moisture loss in the cuttings, and to help cool the upper part of the cutting. The difference in temperature between the top and bottom of the cutting helps encourage growth at the base of the cutting (the warmest part is the most active).

PLANT ESTABLISHMENT METHODS

The hardest part of growing plants is getting them through the first year. After that, they are usually quite well established and can tolerate harsher conditions with little ill effect. Watch for the following when plants are young:

- Drying out (because roots aren't deep into the natural soil where reserves of water can be drawn on hot days).
- Attack by insects or animals eating foliage.
- Attack by insects (e.g. grubs) and animals eating the roots.

- Transplant shock, where the plant, particularly its roots, may be damaged during planting stage or immediately after due to poor environmental conditions (heavy winds, rain).

Never plant during the hottest time of the day. When establishing plants in a garden, it is best to 'nurse' them through the establishment phase. This might involve providing protection and support and then gradually reducing this degree of nurture until the plant is mature enough to survive itself.

Ways to help plants get established

TRICKLE AND DRIP IRRIGATION

Trickle and drip irrigation systems are a simple, cost effective means of irrigation for many gardens. Such systems provide a reliable source of moisture to your plants. This helps both plant establishment and subsequent growth.

Advantages:

- Constant or timed supply of water as desired.
- Reasonably priced and easy to install.
- Parts are usually easy to replace or repair.
- Water use is greatly reduced.
- Water can be supplied to the ground and not to foliage, so there are fewer disease problems (eg. mildews).

Disadvantages:

- Blocked nozzles can be a problem due to residues, or debris (eg. soil) that might build up.
- Nozzles and drippers should be checked to ensure they are still pointing in the desired direction.
- Sometimes moisture on the foliage is desirable.
- Depositing water in one or two spots may not encourage root development in all directions, making the plant less stable.

MULCHES

Mulches assist in plant establishment and subsequent plant growth. Benefits include:

- Reduced competition to your plants by reducing weed growth.
- Conserving moisture in the plant's root zone.
- Organic mulches provide nutrients and humus to the soil as they decompose, thereby aiding plant nutrition and contributing to improved soil structure.
- Mulching acts as a buffer to sudden changes in soil temperature that may damage plant roots. It can be used to provide a layer of insulation between the soil and the hot air above, helping to conserve water and to keep the plant's roots cooler. (This method of keeping roots cool can sometimes allow you to grow some plants in the tropics which would otherwise be difficult in such climates).
- Reduced erosion and compaction of top soil.

Mulches can be composed of inorganic materials (eg. rocks, gravel, synthetics, etc.) or organic materials (eg. newspaper, manures, compost, leaf litter, straw, sawdust, etc.)

Mulches mainly come in two forms:

Mat mulches: flat sheets or layers of materials such as plastic, newspaper, carpet underlay, weed mats, etc.

Bulk mulches: quantities of loose material such as sawdust, barks, leaf litter, grass cuttings, etc.

Bulk mulches are easy to place. They can also be easy to dislodge. Weed growth penetrates bulk mulches more readily than mat types. Bulk mulches tend to be cheaper or more readily available than mat mulches. Some mat types prevent good water penetration (eg. plastic, paper and cardboard). Mat types need to be secured down (ie. pegged to prevent them lifting). The type of mulch you use will depend on factors such as cost, availability, ease of transport and the situation in which it is to be used.

Mulches should be applied while or just after planting. Before mulching, kill or remove any unwanted existing vegetation. This can be carried out by cultivation, hand weeding or by using weedicides. Mulches should be thick enough to provide a good cover of the underlying surface, but not so thick that it will smother your plants. Bulk mulches should be at least 5–7cm thick. If weeds are a problem, 10–15 cm thick would be better. Aim to keep the mulch clear of the plant stems or trunks, particularly organic types of mulch.

Problems with mulches

Mulches that consist of wood or bark products (eg. sawdust, pine bark) will draw nitrogen from the soil unless they are well composted before use. Addition of nitrogen fertilisers will overcome this problem.

- Some mulch materials, particularly those with fine particles, can pack together. This creates a barrier that repels water, thus reducing the amount of moisture to the plant roots. A mixture of mulch materials or composting the material before use will reduce the likelihood of such problems occurring.
- The barks and sawdusts of some trees such as *Pinus* spp. and *Eucalyptus* spp. can contain materials that are toxic to other plants. Be careful if using fresh material derived from these trees or any unknown source.
- Some mulch materials are very light and will easily blow away or be dislodged. Mix such materials with heavier materials to reduce the likelihood of this occurring.
- Slits in mat-type mulches for plantings should be as small as possible and slit edges overlapped to reduce likelihood of weeds growing around the base of the plants.
- Some mulch products may encourage termites, cockroaches or other vermin.

TREE GUARDS

In the tropics, tree guards are used for three main reasons:

- As protection against differing climatic conditions, such as strong winds
- As protection against grazing animals such as rabbits, sheep, cattle, etc.
- As a barrier to prevent damage by machinery, eg. mowers.

Types of tree guards

Plastic tubes

The most durable types are UV stabilised—these should last for several years. In many cases, plant growth is significantly enhanced as the tubes create a warm, moist micro-climate. Protection from frost, wind, and rabbits is excellent. It requires 3-4 stakes to keep the tube upright. Plastic bags, shopping bags or old fertiliser bags can also be used as an alternative, although these are not as long-lasting or effective.

Plastic mesh

Durable guards which are available in both flexible and rigid forms. They require pegs or stakes for support. They don't create a protected, humid environment like plastic tubes.

Plastic pipe

Flexible plastic pipe, 50-100 mm diameter x 500 mm length, can be placed around stems of young frost-sensitive plants (eg. fruit trees). These should be removed after the danger of frost has passed, otherwise fungal problems may occur, as the pipe tends to keep the stem damp.

Wire mesh

Chicken wire tied to stakes or stapled to 3-4 pegs provides a barrier against grazing animals.

Hessian

Hessian bags or cloth need to be tied around 3 stakes. Provides good wind and sun protection, although not in the long term.

Tyres

Old car tyres placed around seedlings can be an effective and cheap barrier against rabbits and hares.

Milk cartons

Useful for marking location of seedlings and providing limited protection against vermin, frost, etc.

COMMON PROBLEMS

The following are some of the most common problems encountered by the home gardener in the tropics and sub-tropics.

- Plant growth and height. Check carefully before purchasing plants. Some subtropical species become invasive when grown in tropical conditions. Some temperate species, if capable of growing in the tropics, may also become weed-like and invasive.

- Weed control. Mulch to control weeds; apply in thick measure for good control. Overuse or misuse of chemicals in a garden can be worse than no chemical control at all, so research your options.
- If plants are doing poorly in a certain area of the garden but quite well everywhere else, look at the past history of that area. There may be chemical or nutrient imbalances that might have occurred by design or mistake. The site may have a slightly different soil type or be prone to flooding, etc.
- Beware of importing pests and diseases into your garden from outside. Buying in soil or sickly plants should be avoided if possible.
- PLAN a garden—this will involve considerations of where to put various species, what species to use and what will be required to care for them.
- Frequent light watering in summer will cause roots of large trees to come to, or develop at, the soil surface. This can lead to damaged paths and buildings and may result in poor drought tolerance. Overwatering can damage plant health, soil structure, nutrient availability, and cause other problems such as fruit splitting, etc.
- Safety with plants. Don't plant large trees too close to power lines or buildings. The maintenance costs of these trees in the future could become very high, and they may eventually pose serious safety problems. Don't choose plants that are known to cause allergies or may be poisonous, particularly if children will visit your garden.
- The high growth rate in the tropics seems to wear the plants out—some plants have a short life-span in the tropics.
- High growth rates require additional watering and fertilisers to keep the plants healthy under those forced growth conditions, compared with similar plants grown in temperate situations.

CONTAINER GROWING AND GARDENING

Pot plants and container gardening can be a great way to add a new dimension to your garden. Consider the following points:

- Plants grown in containers can be moved around for different effects, creating a constantly changing, dynamic garden.
- Some plants look good at certain times of the year, but not at others. When grown in containers, such plants can be shown in a prominent position when they are at their best, and hidden when they don't look their best.
- There may be cases where you want to grow a plant that would not tolerate your soil—the growing medium in your pot will overcome this.
- Vigorous plants can be controlled to some extent by containerising them.
- Many tropical plants look wonderful in pots when brought indoors for special events, particularly when in flower (eg. heliconias).

Points to look out for when growing plants in containers in the tropics:

- Potting media tend to break down more quickly in tropical areas than in temperate climates. This is because most potting mixes have a high organic matter content. As a result, repotting will need to be more frequent.
- Climatic conditions (eg. heat, day length), vigorous growth and quick drainage characteristics of potting media, mean that very regular watering and fertilising is required during active growth seasons.
- Do not use saucers under pots as this may lead to root rots and increased mosquito population (many human diseases are spread by mosquitoes).
- Do not add earthworms to the pots, nor allow other insects or animals to live in the potting mix.

When you go away

In wetter months, you may find placing your containerised plants in a sheltered position is needed. In drier months you may be advised to do one of the following:

- Sit the pot in a trough, bath or other container with water (do not immerse more than the bottom one-third of the soil).
- Install a trickle irrigation system with drippers to each pot (place this on a timer mechanism so it switches on each day for a period).
- Put a slow watering bag on each container plant (ie. a bag filled with water with a dripper tube releasing a very slow drip of water from the bag to the plant).
- Get your neighbours to keep a regular check on your plants. (It is important that your neighbours are reliable, and quite happy to do that for you.)

Problems with containerised plants

- Be sure to choose the right sized pot for the plant. Large plants in small pots can readily fall over, and will struggle to obtain sufficient water and nutrients from the growing medium.
- Choose a growing medium that has sufficient weight to prevent the plant from falling over as it grows larger. Many potting mixes contain lightweight ingredients such as perlite. If there are a lot of such ingredients in the potting mix then there will be insufficient weight in the container to counter-balance the weight of the plant's foliage as it grows.
- If potting mixes dry out too rapidly, then repot with a mix that retains more water, but be careful not to choose one that holds too much moisture.
- Choose containers that have reasonable drainage to prevent waterlogging, but don't dry out too quickly (eg. unglazed ceramics, slatted timber containers, excess drainage holes).
- For plants that are prone to rotting when dormant (eg. some bulbs and herbaceous plants), simply turn containers on their sides to prevent water getting into the containers, or place them in an area where watering is minimal.
- If plants are getting tall and leggy, then the containers may need to be spaced further apart to allow for increased light penetration.

Potting up plants

All plants growing in containers, small or large, need potting up periodically, regardless of the size or type of container. Potting mix will compact over time, making watering more difficult and both drainage and aeration poorer.

Pot up plants as follows:

1. Before removing a plant from an old container for potting up, water it thoroughly. If the plant is dry it is best to immerse the old pot completely in a tub of water. (A wet root ball will come out of an old pot far more easily than a dry one.)
2. If there is a tight mass of roots on the outside of the root ball, you should loosen those roots and break away 10–20% of the old potting mix before repotting.
3. Place some quality potting mix in the bottom of the new container.
4. Place the plant into the new container and fill around it with new potting mix.
5. Water thoroughly, then allow to drain.

2

Trees and Shrubs

The ornamental trees and shrubs dealt with in this chapter can provide a starting point for deciding what might be grown, and how to grow it, but the list is by no means comprehensive.

FAMILY Caprifoliaceae
GENUS *Abelia*
COMMON NAME Abelia, Glossy abelia

Appearance: Spreading flowering shrubs; glossy green leaves.
Establishment: Prefers well drained soil. Feed and water for faster establishment. Grows in sun or shade. Grown from tropics to cooler temperate areas. Propagate by semi-hardwood tip cuttings.
Culture: Prune any time to maintain shape and foliage density. Mulch and irrigate in dry weather.
Cultivars
Approximately 30 species.
A. chinensis—Fragrant white flowers, deciduous, to 1.8m.
A. floribunda—Evergreen, red-purple flowers, to 1.8m.
A. x *grandiflora*—Particularly hardy cultivar. Slender, weeping foliage, white-pink flowers. To 3m tall.
A. schumannii—To 1.8m tall, deciduous, mauve-pink flowers. Better suited to cool areas.
A. uniflora—To 1.8m tall, pale pink flowers with orange throat.

FAMILY Malvaceae
GENUS *Abutilon*
COMMON NAME Indian mallow, Chinese lantern

Appearance: Sprawling shrubs. Crinkled green leaves, some species variegated.
Establishment: Hardy plant, suiting most soils. Full sun or light shade; shelter from strong winds. Grown indoors in colder climates, outdoors in mild temperate to tropical areas. Propagate by semi-hardwood cuttings.
Culture: Prune after flowering to control spread and maintain dense foliage effect. Attacked by leaf-eating insects.
Cultivars
Approx. 150 species.
A. hybridum—To 2m tall, sprawling slender stems. Many cultivars, with varying flower colours (whites, reds, yellows and orange).
A. megapotanicum—Spreading shrub 1–1.8m tall, red flowers.
A. pictum (*A. striatum*)—To 4m. Lobed, palmate leaves, red-veined yellow to orange flowers.

FAMILY Euphorbiaceae
GENUS *Acalypha*
COMMON NAME Copper leaf

Appearance: Mainly medium shrubs with large, colourful leaves. Flowers in long tassels or spikes.
Establishment: Prefers fertile, moist soils and full sun, but generally adaptable. Propagate by cuttings.
Culture: Responds well to watering and mulching during extended dry periods. Lack of sun can result in less colourful foliage. Prune in autumn to shape and promote lush foliage. Can suffer from scale, mealy bug, grass-hoppers.
Cultivars
A. hispida—To 3m tall; long red pendant flower tassels.
A. reptans—Prostrate species suitable for

hanging baskets. Flower spikes of red or pink.

A. wilkesiana—To 3m tall; variable leaf patterns and colours. Small, insignificant flowers.

FAMILY Apocynaceae
GENUS *Adenium*
COMMON NAME Desert rose

Appearance: Small trees or shrubs. Fleshy stems, colourful tubular flowers.
Establishment: Need hot, preferably dry position. Propagate by half-ripe cuttings.
Culture: Full sun to part shade. Prone to mealy bugs and scale.
Cultivars
A. obesum—To 4m tall; deep pink flowers.

FAMILY Mimosaceae
GENUS *Albizzia*
COMMON NAME Powder puff plant

Appearance: Large shrubs and small trees with divided fern-like foliage. Attractive bottlebrush flowers and seed pods.
Establishment: Very easy to grow; some are relatively short-lived. Prefer sandy soil or good drainage. Propagate by seed (soak seed first in boiling water, then remove after water cools).
Culture: Do not prune hard. Hardy, but responds to watering in dry periods.
Cultivars
A. julbrizzin—To 10m, pink flowers. Cold-sensitive.
A. lopantha—Yellow flowers, soft fern-like foliage, to 5m tall.

FAMILY Casuarinaceae
GENUS *Allocasuarina* (*Casuarina*)
COMMON NAME She oak or Casuarina

Appearance: Mainly trees (some large shrubs); needle-like foliage. Tiny, often colourful flowers create a haze of colour over foliage. Similar in appearance to conifers.
Establishment: Most are hardy, adapting to poor conditions. Generally wind- and salt-tolerant. Propagate easily from seed.
Culture: Few pest or disease problems. Seed attracts parrots and other birds. Tolerate pruning for shape only.
Cultivars
The genus *Casuarina* has been split into *Casuarina* and *Allocasuarina*, most species becoming *Allocasuarina*. Some are more suited to tropical conditions than others (eg. *C. equisetifolia* grows throughout the South Pacific tropics)

FAMILY Acanthaceae
GENUS *Aphelandra*
COMMON NAME Zebra plant

Appearance: Dense shrubs or clumps. Foliage has zebra-like markings (various colours) and showy spikes of flower bracts.
Establishment: Prefers moist, drained organic soil, protection from wind, light shade. Avoid low temperatures. Often grown as indoor plant. Propagate by semi-hardwood cuttings.
Culture: Water during dry periods. Prune hard annually to rejuvenate. Easily transplanted. Susceptible to scale and mealy bug.
Cultivars
Approx. 80 species
A. aurantiaca—To 0.8m tall, orange flowers.
A. pectinata—To 0.8m tall, red flowers.
A. sinclairiana—To 5m tall, long reddish flower spikes.
A. squarrosa—To 1m tall, green and white foliage, yellow flowers. Commonly grown species.
A. tetragona—To 0.8m, scarlet flowers.

FAMILY Myrsinaceae
GENUS *Ardisia*
COMMON NAME Spear flower

Appearance: Trees and shrubs. Clusters of small flowers usually followed by attractive berries
Establishment: Prefer filtered sun or shade in hottest part of the day. Like drained soil high in organic matter. Grow well from tropics to mild temperate climates. Propagate by seed sown 5mm deep, or heel cuttings in spring kept at 24°C.
Culture: Water in extended dry periods. Remove any weak or straggly shoots after berries drop.
Cultivars
A. crenata—White flowers followed by red berries, to 1m tall.
A. crenata var. *alba*—White berries.

A. esculenta—To 0.7m tall, pale purple berries. Best in tropics.

A. glandulosa var. *marginata*—To 7m tall, shrubby tree. Crimson berries turn black. Suited to wet tropics.

A. macrocarpa—To 1m or taller, large bright red berries.

FAMILY Acanthaceae
GENUS *Barleria*
COMMON NAME Barleria

Appearance: Flowering, upright shrubs and herbs, with slender branches; blue or white flower clusters.

Establishment: Suit wet tropics, shade. Plant after cold weather. Suitable for hedging. Propagate by softwood cuttings kept at 29°C.

Culture: Water lightly in winter, well in warm weather. Occasionally attacked by thrip and grasshoppers.

Cultivars

Approx. 230 species

B. cristata—To 1m, purple or white flowers.

B. flava To 2m tall, yellow flowers.

B. involucrata—To 70cm tall, blue flowers.

B. lupulina—To 0.7m, spiny, yellow flowers.

B. stigosa syn. *B. caerulea*—To 0.9m tall, blue flowers.

FAMILY Caesalpiniaceae
GENUS *Bauhinia*
COMMON NAME Orchid tree

Appearance: Mainly flowering shrubs and trees and some low, sprawling plants.

Establishment: Suit tropics and sub-tropics, occasionally extending to milder climates. Prefer full sun, adapt to most soils. Respond to watering and feeding. Propagate by cuttings kept at 24°C, under moist or humid conditions, also by seed or layering.

Culture: Water during drought. Prune for shape only. Few pests.

Cultivars

Around 300 species.

B. galpini—To 1.8m tall, white summer flowers.

B. grandiflora—To 1.8m tall, white flowers.

B. natalensis—White autumn flowers.

B. purpurea—To 5m tall, red and white flowers.

B. tomentosa—To 5m tall, yellow and red flowers.

B. variegata—To 7m tall, variegated leaves, red and yellow flowers.

FAMILY Bixaceae
GENUS *Bixa*
COMMON NAME Lipstick tree

Appearance: Flowering bushy tree. Shiny, green leaves contrasting with dark stems. Pink flowers in summer followed by decorative, reddish seed capsules.

Establishment: Prefers fertile soil and full sun. Propagate by cuttings, shoots 6–12 months old, kept at 30°C.

Culture: Water in dry periods. Can be pruned to dense bush. Relatively pest-free.

Cultivars

One species, *B. orellana*—small evergreen tree, heart-shaped leaves, decorative seed capsules.

FAMILY Sterculiaceae
GENUS *Brachychiton*
COMMON NAME Flame tree, Kurrajong

Appearance: Medium-sized trees or shrubs, some deciduous, often with thickened trunks. Clusters of colourful bell-shaped flowers, seed in pods. (Similar to *Sterculia*, but with hairy inside on seed capsule.)

Establishment: Most prefer open position, full sun. Many are drought- and wind-hardy, some hardy into temperate climates. Propagate by seed; cuttings are also possible.

Culture: Few problems once established.

Cultivars

B. acerifolius—Grows 15–30m tall, bright red flowers.

B. australis—Blue-green foliage.

B.bidwillii—To 4m tall, from coastal Queensland. Pink flowers.

B.bicolor—To 30m, deciduous in cooler climates. Red to pink flowers.

B. gregorii (Desert Kurrajong)—Yellow and red flowers.

B. populneus— Grows 10–20m tall, large poplar-like leaves. Flowers pale outside and red to yellow inside.

FAMILY Euphorbiaceae
GENUS *Breynia*
COMMON NAME Confetti plant, Zig Zag plant, Snow bush

(NB: Differs from *Phyllanthus* in fine detail of flower structure only.)

Appearance: Shrubs and trees, some with attractive foliage, berry fruits.

Establishment: Require warm to hot climates; best in ordinary to sandy soil and full sun. Propagate by cuttings or suckers.

Culture: Water in dry periods. Fertilise to promote healthy foliage. Prone to defoliation by beetles. Prune to maintain shape.

Cultivars

B. disticha—To 1.5m tall, reddish concertina-shaped stems.

B. disticha 'Rosea'—Red/pink variegated leaves.

FAMILY Solanaceae
GENUS *Browallia*
COMMON NAME Bush violet

Appearance: Small herbaceous plants with attractive flowers.

Establishment: Tropics to sub-tropics. Requires protection from frost; best in part shade. Propagate by cuttings or seed.

Culture: Water in dry periods. Fertilise regularly to encourage growth. Prune to maintain shape.

Cultivars

B. americana—To 0.7m, mauve flowers with yellow centre.

B. speciosa—To 1.6m tall. Several varieties with flower colours varying from blues to whites.

B. viscosa—To 30cm, prolific flowering. Flower colours according to variety.

FAMILY Solanaceae
GENUS *Brunfelsia*
COMMON NAME Yesterday today and tomorrow

Appearance: Evergreen, flowering shrubs or small trees. Fragrant phlox-like flowers changing colour with age.

Establishment: Best in rich, fertile organic soil, adaptable to other soils. Frost-tender. Propagate by 6cm cuttings kept at around 18–21°C.

Culture: Respond well to mulch and feeding. May need irrigation during dry periods.

Cultivars

B. americana—To 5m tall, yellow to white summer flowers.

B. australis—Grows 3–4m tall. Purple and white fragrant flowers fade to white.

B. latifolia—To 1m tall. Purplish or white flowers, winter or spring.

B. pauciflora syn. *B. calycina*—To 0.7m tall. Fragrant purple and white summer flowers.

FAMILY Rubiaceae
GENUS *Bouvardia*
COMMON NAME Bouvardia

Appearance: Mainly sprawling, tender, woody shrubs and herbs, with showy flower clusters, sometimes fragrant.

Establishment: Prefer well drained and ventilated site; adapt to most soils. Thrive at temperatures from around 18 to 25°C. Not suited to prolonged heat and humidity of the wet tropics. Frequent tip pruning of young plants needed to develop more compact foliage. Propagate from semi-hardwood cuttings.

Culture: Prune back very hard after flowering (50–90% annually). Water well in dry periods. Respond to regular fertilising. Few serious pests.

Cultivars

B. leiantha—Deep red flowers.

B. humboldtii (scented) syn. *B. longiflora*—To 1.7m tall, fragrant white flowers.

B. longiflora 'President Cleveland'—To 1m, red flowers.

B. ternifolia—To 2m, red flowers.

FAMILY Caesalpiniaceae
GENUS *Caesalpinia*
COMMON NAME Poinciana, Leopardwood

Appearance: Trees, shrubs and occasionally climbers. Clusters of red or yellow flowers, divided leaves. Fruit a flattened pod.

Establishment: Tropical or subtropical; some may adapt to warmer temperate sites. Prefer fertile, moist sandy soils. Protect from strong winds. Propagate by cuttings under moist, base heat between 25 and 30°C. Seed will germinate if soaked in warm water for a few hours before sowing.

Culture: Relatively hardy once established. Water during dry periods.

Cultivars

C. ferrea—Small or large tree, sometimes to 15m tall.

C. gilliesii (Bird of paradise tree)—Shrubby tree to 4m+, yellow flowers with red stamens.

C. japonica—To 3m+, yellow flowers and red stamens in loose clusters.

FAMILY Mimosaceae
GENUS *Calliandra*
COMMON NAME Powder puff

Appearance: Shrubs and trees with fine feathery foliage and attractive ball-shaped flowers, flowering much of the year in the tropics. Leaves are light-sensitive, closing at night.

Establishment: Suited to wet tropics or sub-tropics, may extend to milder areas. Prefer full sun and good drainage. Propagate by cuttings over heat at around 25°C.

Culture: Prune to contain size. Few pests or diseases, generally long-lived. Atracts birds.

Cultivars

C. haematocephala—To 4m tall, red flowers.

C. portoricensis—To 5m+, white flowers, dark green foliage.

C. pulcherrima syn. *C. tweedii*—To 2m+ tall and 2m diameter, red flowers.

C. surinamensis—To 4m tall, sparse leaf cover, white flowers with pink tips.

FAMILY Verbenaceae
GENUS *Callicarpa*
COMMON NAME Beautyberry

Appearance: Deciduous and evergreen shrubs and trees, from tropics and sub-tropics, both hardy and tender species. Produces masses of fruits.

Establishment: Prefer good ventilation, well lit position. Propagate by cuttings kept at around 27°C, by layers or seed.

Culture: Pinch back tips of long shoots to encourage dense branching. Prune straggly shoots after fruiting or before potting to encourage new fruit-bearing shoots.

Cultivars

C. americana—To 2m. Violet fruits, bluish flowers.

C. dichotoma (*C. purpurea*)—Decorative shrub – 2m tall. Dense mass of fine stems bearing clusters of tiny mauve berries.

C. japonica—To 1.3m, violet fruits.

C. longifolia—To 5m, white or deep pink fruits.

C. nudiflora—To 9m, blue fruits, tiny red or purple flowers.

FAMILY Myrtaceae
GENUS *Callistemon*
COMMON NAME Bottlebrush

Appearance: Shrubs or small trees. Bottlebrush flowers more than 5cm in diameter.

Establishment: Best planted young. Protect from wind, heat and dryness for first 12 months, after which they are generally hardy. Seed germinates readily within 5–10 days of sowing.

Culture: Hardy plants, cold- and heat-tolerant; some tolerate high humidity. Appreciate mulch and water in dry periods. Prune lightly and often for shape.

Cultivars

C. citrinus—To 6m tall, lemon-scented leaves.

C. viminalis—To 7m+, red brushes. Grows well in tropics. Numerous varieties.

FAMILY Theaceae
GENUS *Camellia*
COMMON NAME Camellia, Tea plant

Appearance: Evergreen shrubs and trees; glossy, dark green foliage

Establishment: Prefer semi-shade, protection from wind, in fertile, well drained mulched and acidic soil. Plant after summer heat. Propagate by cuttings in a hot bed.

Culture: Like regular watering and fertilising. Prune for shape. Can suffer from scale, aphids, mealy bug, nematodes and leaf-eating insects. Sunburn can cause brown patches on leaves.

Cultivars

Suitable for tropics and sub-tropics:

C. sasanqua—To 2.5m+. Smaller leaves than *C. japonica*. Lots of varieties, with many flower colours.

C. sinensis (Commercial tea)—Grows 1–15m, depending on variety.

Family Annonaceae
Genus *Cananga* syn. *Canangium*
Common Name Cananga, Ylang Ylang

Appearance: Trees with long, arching semi-pendulous branches. Scented, yellow flowers with star-shaped petals.
Establishment: Suited to tropical climates only.
Culture: Moist, fertile soil preferred.
Cultivars
C. odorata var. *macrophylla* syn. *C. odoratum*—To 27m, fragrant, yellow-green flowers. Slight pepper smell from crushed leaves. Flowers are source of cananga oil.
C. odorata var. *genuina* (Ylang Ylang)—To 15m, flowers and leaves smaller than var. *macrophylla*. Flowers are source of Ylang Ylang oil.

Family Apocynaceae
Genus *Carissa*
Common Name Natal plum

Appearance: Prickly shrubs or small trees with dense network of thin branches and leathery leaves. Clusters of fragrant pink or white flowers, attractive berry fruit.
Establishment: Best in full sun. Grown for edible fruits, or hedges. Propagate by seed or cuttings.
Culture: Respond well to watering and mulching when in growth. Fertilise annually. Prune for shape and to control size.
Cultivars
C. bispinosa—To 3m, white flowers.
C. karandas (Karanda)—Shrub or small tree, white or pink flowers.
C. edulis—To 3m tall, sprawling shrub. Flowers white with mauve or reddish inside; red to reddish-purple fruits.
C. macrocarpa (Natal plum)—To 5m tall, white flowers, red fruits to 6cm diameter. Several other forms of this species.

Family Caesalpiniaceae
Genus *Cassia*
Common Name Cassia or Senna

Note: Cassia has been split into *Senna* and *Cassia*. *Senna* has flowers emerging from the sides of shoots (ie. axillary) and flattened pods; *Cassia* has flowers at tips (ie. terminal) and cylindrical pods.
Appearance: Mainly shrubs with yellow flowers and pods that are round in cross section.
Establishment: Well drained moist soil, sunny location. Propagate by seed that has been immersed in boiling water, then removed and soaked in cool water for a few hours.
Culture: Withstand light frost. Prune lightly to shape.
Cultivars
C.artemisioides—Rounded shrub to 2m, silver leaves, yellow flowers.
C.odorata syn. *sturtii*—Grows to 2 m.

Family Solanaceae
Genus *Cestrum* syn. *Habromanthus*
Common Name Jessamine

Appearance: Shrubs and small trees, often sprawling, with multiple long stems. Attractive clusters of tubular flowers. Crushed foliage often fragrant.
Establishment: Many will grow in mild temperate climates, but generally do better in sub-tropics. Propagate by semi-hardwood cuttings, kept around 24°C.
Culture: Prune for shape and to promote flowering tips.
Cultivars
C. aurantiacum—Grow 2-5m tall, scented yellow to orange flowers.
C. elegans syn. *C. purpureum*—To 3m tall, reddish to purple flowers.
C. fasciculatum—To 3m tall, mauve flowers.
C. newellii—To 3m tall, bright crimson flowers.
C. nocturnum (Night-scented Jessamine)—To 3.5m tall. Cream flowers heavily scented at night.
C. parqui—Around 5m tall. Pale-green to brownish flowers, long narrow leaves toxic to animals.

Family Acanthaceae
Genus *Crossandra*
Common Name Firecracker flower

Appearance: Shrubs or herbs, ovate leaves. Attractive flowers clustered in terminal spikes.
Establishment: Do well in freely draining

organic soil. Propagate by 6cm cuttings kept at 27–30°C.

Culture: Reduce watering over winter but keep soil moist when putting on growth. Respond to feeding.

Cultivars

C. guineensis—To 40cm tall, lilac flowers.

C. infundibuliformis—To 1m tall, orange to red flowers.

C. mucronata—To 1m tall, red flowers.

FAMILY Euphorbiaceae
GENUS *Codiaeum*
COMMON NAME Croton

Appearance: Medium-sized shrub. Large, often multi-coloured leathery leaves, sometimes lobed.

Establishment: Grows in any reasonable soil. Avoid poorly drained sites. Full sun for best foliage colour. Propagate by semi-hardwood cuttings or aerial layers.

Culture: Prune for shape and to rejuvenate old wood. Susceptible to scale and mealy bug, which can lead to sooty mould.

Cultivars

C. variegatum—To 2m, variegated leaves of red, white and yellow. Large range of cultivars.

FAMILY Caesalpiniaceae
GENUS *Delonix*
COMMON NAME Poinciana

Appearance: Trees with divided, feathery leaves and attractive flowers.

Establishment: Prefers well drained friable soil, frost-free area. Propagate by stem cuttings placed under heat and mist (or fog).

Culture: Reduce watering in winter, keep moist during warmer weather.

Cultivars

D. elata—Smaller flowers than *D. regia*. Pale-yellow flowers age to reddish colourings.

D. regia (Royal poinciana)—To 14m, red flowers. Most widely grown species.

FAMILY Araliaceae
GENUS *Dizygotheca*
COMMON NAME False aralia

Appearance: Evergreen shrubs and small trees. Palmately compound leaves.

Establishment: Native to Pacific islands. Like reasonable drainage; appreciate fertile soil in protected, partly shaded position. Grow well in containers. Propagate by semi-hardwood or hardwood cuttings over heat, or by root cuttings, or grafting.

Culture: Water during dry periods.

Cultivars

D. elegantissima—To 3m+. Long, narrow, drooping leaflets, reddish brown in colour.

D. veitchii—Narrow leaflets, green on top with coppery tones underneath.

FAMILY Verbenaceae
GENUS *Duranta*
COMMON NAME Pigeon berry

Appearance: Shrubs and trees, sometimes prickly. Clusters of simple white or purplish flowers.

Establishment: Grow well in full sun; prefer moist, drained soil. Generally hardy. Propagate by cuttings or seed.

Culture: Can be pruned regularly for shape; suitable as a hedge.

Cultivars

D. lorentzii—To 3m tall, white flowers.

D. repens syn. *D. plumieri*—Medium shrub to 6m tree. Foliage colour golden to silvery or white leaves, most covered with prickles. White or purple perfumed flowers.

D. stenostachya—Small spineless tree to 5m tall, lilac flowers.

FAMILY Boraginaceae
GENUS *Echium*
COMMON NAME Viper's bugloss

Appearance: Herbs or shrubs; dense, often hairy foliage. Colourful flowers develop on large erect spikes above the foliage.

Establishment: Generally hardy, adaptable to both cooler and warmer areas. Best in full sun. Keep roots moist when first planted. May need staking. Propagate easily by seed, cuttings or layering.

Culture: Protect from wind and frosts. Water during extended dry periods. Keep roots mulched.

Cultivars

E. candicans—To 1.8m; white, hairy foliage, blue or white flowers.

E. creticum—To 1m tall; purplish to reddish flowers changing colour with age.

E. fastuosum—To 2m tall; greyish, hairy foliage, purple to blue or white flowers.

E. vulgare—To 1m tall. Usually blue flowers, sometimes white or pink. A weed in some places.

FAMILY Elaecarpaceae
GENUS *Elaeocarpus*
COMMON NAME Blueberry ash

Appearance: Rainforest trees with simple, glossy leaves. Clusters of small decorative berries attract birds. Bell-shaped flowers with serrated margins.

Establishment: Prefers filtered light to sunny position and friable organic soil. Propagate by tip cuttings, or fresh seed.

Culture: Mulch and irrigate during dry periods. Few serious pests. Foliage can become sparse in shaded situations. Prune back after fruiting to promote new wood and keep foliage dense.

Cultivars

E. dentatus—To 20m tall. Silky leaves mainly at tips, purple-grey fruit clusters.

E. reticulatus—Grows 10–20m tall, blue fruits. Grows from subtropical to temperate Australia.

E. sphaericus—Tree, purple fruits. Native of India.

FAMILY Acanthaceae
GENUS *Eranthemum*
COMMON NAME Blue sage

Appearance: Evergreen herbs and shrubs. Dense, attractive foliage with prominent veins. Some species have coloured foliage, tubular flowers.

Establishment: Prefer friable, sandy loam, mulching, and filtered sunlight or partial shade. Propagate by cuttings.

Culture: Water during dry periods. Direct sun burns foliage. Susceptible to mealy bug.

Cultivars

E. pulchellum (Blue sage)—1m tall, woody stems, green foliage, deep-blue to purple flowers.

E. tricolor syn. *Pseuderanthemum tricolor*—Spreading shrub; leaves blotched with pink or purple.

FAMILY Fabaceae
GENUS *Erythrina*
COMMON NAME Coral tree

Appearance: Mainly deciduous trees and shrubs, some evergreen, some herbs. Foliage often sparse, only at growing tips. Many species flower when bare of foliage. Flowers generally brilliant, reddish tones.

Establishment: Best suited to climates with distinct seasons. Prefer well drained friable, fertile soil, although some will adapt readily to most soils. Full sun or filtered sunlight (few like heavy shade). Some frost-tender, particularly when young. Respond to watering and feeding when first planted. Many will grow in wet tropics and drier inland areas. Propagate by semi-hardwood, hardwood and soft cuttings with a heel, or seed.

Culture: Most tolerate hot and dry periods once established. Few serious pests or diseases. Prune for shape.

Cultivars

E. abyssinica—To 12m tall; dark cork-like bark, red flowers.

E. acanthocarpa—To 2m tall. Prickly branches, red and yellow flowers on bare drooping branches in early spring, before leaves emerge.

E. berteroana—To 7m+. Flowers when bare of leaves. Decorative contorted seed pods to 30cm long.

E. coralloides—Prickly tree to 7m.

E. crista-galli—Tree 3–5m+ tall, scarlet flowers.

E. fusca syn. *E. glauca, E. ovalifolia*—Tree to 10m+. Brownish or beige flowers, leathery foliage.

E. poepppigiana—To 20m tall, prickly, orange flowers.

E. subumbrans—To 15m+. From tropical south-east Asia.

E. tahitensis syn. *E. sandwiciensis*—To 10m tall. Leafless when flowering, reddish orange flowers.

E. variegata syn. *E. indica*—To 20m tall, variegated leaves. Widely cultivated.

FAMILY Apocynaceae
GENUS *Ervatamia* syn. *Tabernaemontana*
COMMON NAME False gardenia, Wax flower

Appearance: Shrubs or trees. Shiny, green foliage; tubular flowers similar to gardenias, milky sap.
Establishment: Need full sun. Tolerate hot dry conditions, do well in wet tropics.
Culture: Prune for shape. Can suffer from various insects and sooty mould. Respond to feeding and watering in extended dry conditions.
Cultivars
E. coronaria—To 3m, waxy white fragrant flowers.
E. grandiflora—To 3m, bright yellow flowers.

FAMILY Myrtaceae
GENUS *Eucalyptus*
COMMON NAME Gum tree

Appearance: Small to large trees, occasionally multi-stemmed trunk. Characteristic gum scent in foliage; clusters of colourful flowers.
Establishment: Best planted young; advanced specimens have difficulty developing strong roots and may be susceptible to blowing over in strong winds. Great variation between species, so it is important to select the appropriate species for your site. Propagate easily by seed.
Culture: Prune young plants to shape and encourage one main trunk. Respond to feeding, mulching and watering in dry periods. Very hardy once established. Insects may chew leaves or bore in wood, but a healthy tree will survive such problems.
Cultivars
Over 500 species. Many suit dry tropics and wet tropics; some only suited to temperate climates.
E. camaldulensis (River Red Gum)—Large hardy tree, attractive white blotched trunk. Tolerant of flood and drought.
E. citriodora—To 30m+, lemon scented foliage, attractive salmon-pink bark.
E. ficifolia—Small tree; showy flowers red, orange or white.
E. howittiana—To 20m+, cream flowers, rough bark.
E. papuana (Ghost gum)—To 16m, smooth white bark.
E. ptychocarpa—To 10m, white, pink or red flower clusters.
E. tessellaris—To 20m+, slender weeping leaves.
E. ptychocarpa X *ficifolia* hybrids—New varieties bred in south-east Queensland, with particularly colourful flowers, and suited to warm, humid climates.

FAMILY Euphorbiaceae
GENUS *Euphorbia*
COMMON NAME Spurge

Appearance: Variable herbs, trees and shrubs. All have skin-irritating milky sap. Some are prickly, some look like succulents, others appear more woody. Flowers are generally insignificant but are surrounded by colourful bracts occurring in umbrella-like clusters.
Establishment: Need good drainage, protection from frost. Prefer full sun or at least good levels of light; tolerate heat and dryness. Respond to feeding and moist (not saturated) soil, particularly while establishing. Most propagate easily from cuttings almost any time.
Culture: Prune for shape and to rejuvenate. Susceptible to root rots. Toxins from sap can poison or irritate animals, including fish— be careful.
Cultivars
Over 1600 species and many cultivars.
E. pulcherrima (Poinsettia)—One of the most widely grown species, growing to 3m tall, with large green leaves and large bright red bracts. Other cultivars are available with white, pink and other coloured bracts. Grown as an indoor plant in both temperate and warm climates, and as an outdoor shrub in frost-free climates.

FAMILY Moraceae
GENUS *Ficus*
COMMON NAME Fig

Appearance: Mainly trees, some smaller plants. Generally thick, leathery, shiny leaves, and milky sap.
Establishment: Mainly wet tropical and subtropical climates, but some tolerate dryness and cold, particularly once established. Protect young plants from extremes. Adapt to most

soils, but appreciate moisture and fertility. Roots can be invasive, damaging constructions. Propagate by seed; will also grow from cuttings or aerial layers.

Culture: Generally very easy to grow once established, provided they do not suffer extended periods of dryness. Prune roots and foliage to control size. Caterpillars sometimes attack trees.

Cultivars

Approx 800 species.

F. benjamina—Small tree, attractive foliage.

F. elastica—To 10m, dark leathery leaves.

F. macrophylla—Moreton Bay fig—To 20m+, broad, spreading, dark leaves.

F. rubiginosa syn. *F. australis*—Small tree, rusty new growth.

FAMILY Malphighaceae
GENUS *Galphimia* syn. *Thryallis*
COMMON NAME Galphimia

Appearance: Thin-stemmed shrubs, spikes of yellow star-like flowers.

Establishment: Hardy and fast growing. Prefers friable soil, full sun. Propagate by cuttings or seed.

Culture: Easily transplanted. Lightly prune annually.

Cultivars

G. glauca—To 2.5m, grey-green foliage. Yellow pea flowers all year in tropics. Reddish hairs on stems.

G. gracilis—To 3m.

FAMILY Rubiaceae
GENUS *Gardenia*
COMMON NAME Gardenia

Appearance: Shrubs and small trees. Glossy, dark leaves; large, solitary white or yellowish flowers, sometimes scented.

Establishment: Need friable, well drained, moist organic soil. Need protection from extreme heat, cold or wind. Ideal in filtered sunlight. The species below have also been grown successfully in temperate climates. Propagate by cuttings kept at 24–28°C.

Culture: Respond well to feeding, watering in dry periods, and mulching. Avoid waterlogging. Iron deficiency sometimes causes yellowing of tips. Prune for shape and rejuvenation after flowering. Sometimes suffer scale, mealy bug, sooty mould, nematodes and chewing insects.

Cultivars

Around 200 species. The most commonly grown ones include:

G. jasminoides syn. *G. florida, G. grandiflora*—There are many varieties of this species, some up to 2m tall. Perfumed flowers, usually doubles (sometimes single) and up to 7cm diameter.

G. radicans—A smaller spreading shrub, no more than 0.5m tall. Prolific scented flowers.

G. thunbergia—To 4m tall. Large, scented flowers atop a long floral tube. Large and decorative woody fruits.

FAMILY Verbenaceae
GENUS *Gmelina*
COMMON NAME Gmelina

Appearance: Often prickly, small trees and shrubs, yellow flowers.

Establishment: Prefer well drained, fertile, moist soil. Avoid frosts. Protect young plants from wind. Propagate by cuttings or seed.

Culture: Water often in dry periods. Respond to regular feeding. Lightly prune after flowering.

Cultivars

G. arborea—To 20m, deciduous, yellow flowers.

G. asiatica—Spiny shrub, yellow flowers.

FAMILY Acanthaceae
GENUS *Graptophyllum*
COMMON NAME Caricature plant

Appearance: Similar to *Acalypha*. Shrubs or small trees, often with large colourful leaves.

Establishment: Prefer fertile, moist soils and full sun, but generally adaptable. Propagate by cuttings.

Culture: Respond to watering and mulching during extended dry periods. Lack of sun can result in less colourful foliage. Prune in autumn to shape and promote lush foliage. Can suffer from scale, mealy bug, grasshoppers.

Cultivars

G. pictum syn. *G. hortense*—To 2.5m tall. Dark purplish-green foliage with yellow markings, reddish or purple flowers.

FAMILY Proteaceae
GENUS *Grevillea*
COMMON NAME Spider flower

Appearance: Highly variable shrubs, some trees and ground covers. Dense clusters of attractive flowers, with prominent stigmas.

Establishment: Some species suit hot climates, others only temperate areas. Prefer well drained soils, tolerant of range of conditions. Avoid lime soils. Propagate by cuttings, seed propagation. Some may be grafted.

Culture: Tip prune; heavy pruning can result in die-back. Mulch, water only in drought. Few serious pests or diseases. Suffers root rots.

Cultivars

Sub-tropics and tropics species are mainly hybrids. Many have *Grevillea banksii* as one of their parents. Varieties of the *G.banksii* hybrids include:

G. 'Caloundra Gem'—*G. banksii* x *G.* 'Honeycomb'. 3–5m tall, 3m diameter. Pink and yellow flowers 20cm long, all year. Prefers sunny position.

G. 'Coconut Ice'—*G. banksii* x *G. bipinnatifida*. 1.5m tall, 1.5m diameter. Pink and orange flowers 20cm long, all year. Prefers full sun. Very hardy. Prune regularly.

G. 'Honeycomb'—form of *G. whiteana*. 3–4m tall, 2m diameter. Creamy yellow flowers to 12cm long in winter and spring. Relatively frost hardy.

G. 'Honey Gem'—*G. banksii* x *G. pteridifolia*. Orange flowers with some pink, 20cm long. Grows 4m tall and 2.5m diameter. Very fast-growing, hardy.

G. *banksii* 'Kingaroy Slippers'—Upright form.

G. 'Majestic'—hybrid of G. 'Pink Surprise'; 4m tall, 2m diameter. Flowers red with cream styles, 30cm long, all year. Reasonably hardy but needs full sun.

G. 'Masons Hybrid' syn. G. 'Ned Kelly'—hybrid of *G. banksii* x *G. bipinnatifida* (orange form). 2m X 1.5m. Covered in orange and red flowers all year. Very hardy, tolerates frost, grows well in most soils.

G. 'Misty Pink'—*G. banksii* x *G. sessilis*. Hardy, vigorous, greyish leaves, 3m tall X 2m diameter. Flowers are pink with cream styles, blooming all year. Prune regularly and remove seed pods after flowering to increase flowers.

G. 'Moonlight'—*G. whiteana* hybrid. A rapid-growing shrub to 4m tall and 1.5m across. Large white flowers all year. Prefers full sun and good drainage. Prune regularly to keep bushy.

G. 'Pink Parfait'—G. 'Misty Pink' hybrid. Upright plant to 3.5m and 1.5m diameter. Leaves are silvery coloured. Flowers are watermelon pink with gold tips, to 18cm long. This plant is frost-tender, but otherwise hardy. Regular pruning encourages flowering.

G. 'Pink Surprise'—*G. banksii* x *G. whiteana*. Pink flowers with cream styles to 22cm long. Flowers all year. Vigorous, frost-hardy, but susceptible to wind damage. Prune regularly . Grows 3–6m tall and 2m diameter.

G. *banksii* 'Prostrate alba'—Low growing, white flowered form.

G. 'Robyn Gordon'—To 1.5m tall and 2m diameter. A very popular plant throughout Australia. Hardy and adaptable, best in open, sunny, well drained position. Slightly pendulous red flowers all year. Responds well to pruning. Is susceptible to sooty mould and leaf spotting fungi.

G. *banksii* 'Ruby Red'—Smaller bright red flowers, dense foliage.

G. 'Sandra Gordon'—*G. sessilis* x *G. pteridifolia*. Yellow flowers to 20cm long, except for summer season. Prune after flowering in winter or spring. Susceptible to wind, frost-hardy.

G. 'Starfire'—G. 'Honey Gem' hybrid. To 4m X 2m. Flowers dark red striped centre and pink styles tipped with gold. Flowers all year. Best pruned.

G. 'Sylvia'—G. 'Pink Surprise' hybrid. Soft silvery foliage. Grows to 3 X 2m. Tolerates light frosts. Flowers bright pink with gold tips and to 25cm long. Prune to keep in shape.

G. 'Superb'—*G. bipinnatifida* x *G. banksii* 'Alba'. Dense plant to 1.5m X 1.5m. Orange and pink flowers tipped with gold. Very hardy. Regular pruning encourages better flowering.

FAMILY Tiliaceae
GENUS *Grewia*
COMMON NAME Lavender star

Appearance: Rambling shrub or climber, with yellow to purplish star-shaped flowers.
Establishment: Likes fertile soil. Good hedge plant. Protect from cold. Propagate from hardwood cuttings.
Culture: Fast-growing. Few serious pests, but subject to scale.
Cultivars
G. caffra—Spreads to 3m. Lavender-pink flower, yellow stamens.
G. occidentalis—To 3m. Mauve pink flower with pink/purple inside.

FAMILY Malvaceae
GENUS *Hibiscus*
COMMON NAME Hibiscus, Rose mallow
Appearance: Large shrubs to trees. Showy, single flowers in shades of whites, reds and purples.
Establishment: Prefers good drainage, full sun, protection from cold and dry conditions. In exposed conditions, establish with a tree guard. In protected gardens, hibiscus do well as container plants, or as a hedge or standard. Commonly propagated by cuttings to ensure dependable flower characteristics.
Culture: Responds to heavy pruning after flowering. Susceptible to sucking insects and soot mould. Avoid waterlogging. Feed regularly with rotted manure or balanced liquid fertiliser.
Cultivars
Species number 250, many from tropics and sub-tropics. Three main groups most widely grown in the tropics:
H. rosa-sinensis—Hawaiian hibiscus—Large colourful flowers with crinkled petals. Most commonly grown group, including hundreds of varieties.
Fijian hibiscus—Hardier than Hawaiian. Easier to cultivate in harsh conditions, more suited to exposed sites.
Clark cultivars—Bred from Hawaiians and Fijians. Better colours than Fijians and hardier than Hawaiians.

FAMILY Verbenaceae
GENUS *Holmskioldia*
COMMON NAME Hat plant
Appearance: Woody sprawling shrub. Sometimes prickly, simple leaves. Clusters of colourful flowers.
Establishment: Needs warm position, sun or semi-shade, good drainage. Propagate by cuttings.
Culture:
Relatively pest-free. Prune often to keep foliage dense and bushy.
Cultivars
H. sanguinea—To 4m+. Scarlet and orange flowers, deciduous.
H. sanguinea 'Aurea'—Rambling plant, orange-yellow flowers.

FAMILY Solanaceae
GENUS *Iochroma*
COMMON NAME Iochroma
Appearance: Attractive shrubs, often with soft hairy or downy foliage and colourful bell-shaped flowers.
Establishment: Cultivated varieties mainly from tropical America. Prefer humid, hot conditions. Provide some protection from extremes such as strong wind or cold. Propagate by cutting or seed.
Culture: Water and mulch in dry periods. Propagate by hardwood cuttings.
Cultivars
I. coccineaum—Red flowers.
I. lanceolatum—Dark blue to purple flowers.
I. tubulosum—2m, grey green foliage, blue to purple flowers.
I. warscewiczii—Spreading shrub. Large soft leaves, lavender to blue flowers with dark stripes.

FAMILY Rubiaceae
GENUS *Ixora*
COMMON NAME Flame flower
Appearance: Shrubs and trees, with attractive foliage and colourful ball flower heads, sometimes fragrant.
Establishment: Prefer minimum night temperatures of 15–20°C, though many tolerate lower. Like full sun, acidic, moist well-drained soil. Propagate by cuttings kept at 25–30°C.
Culture: Mulch and feed well rotted manure annually; water during dry periods. Prune lightly after flowering.

Cultivars

I. chinensis—Grows 1–2m, many cultivated varieties.

I. coccinea—Up to 3m tall, various colours.

FAMILY Bignoniaceae
GENUS *Jacaranda*
COMMON NAME Jacaranda

Appearance: Shrubs and trees with soft fern-like foliage. Bunches of flowers commonly blue to purple but occasionally pink or white.

Establishment: Tropical, though some can be grown in temperate climates. Need reasonable drainage, fertile moist soil, protection from frost. Open position in full sun. Tolerant of dry or humid climates. Propagate by seed or semi-hardwood cuttings.

Culture: Prune young plants freely to shape. Few pests and diseases; occasionally mites occur.

Cultivars

J. mimosifolia—Grows 10–15m tall. Most commonly cultivated species.

GENUS *Jacobinia*—see *Justicia*

FAMILY Euphorbiaceae
GENUS *Jatropha*
COMMON NAME Coral plant

Appearance: Shrubs, herbs and trees. Flowers purplish, red or yellow.

Establishment: Light to medium shade will suit most species. Fertile, moist but drained soil is preferred. Seeds can be toxic. Propagate by seed or cuttings.

Culture: Like regular feeding and shelter from strong winds. Avoid waterlogging, but water during dry periods.

Cultivars

J. curcas—To 4m tall, yellow flowers.

J. integerrina—Evergreen shrub to 10m, bright red flower clusters.

J. multifida—To 20m (some forms much smaller). Deeply divided leaves, scarlet flowers.

FAMILY Acanthaceae
GENUS *Justicia*—syn. *Jacobinia*
COMMON NAME Shrimp plant

Appearance: Small shrubs with showy, weeping bracts. Flowers white to purples and pinks.

Establishment: Rich soil, shaded position. Can become a weed problem. Propagate by cuttings or seed.

Culture: Prune regularly to control growth. Responds well to feeding and watering.

Cultivars

J. obtusior—Small shrub, pink or crimson flowers.

J. pauciflora—Small shrub, profuse scarlet and yellow flowers.

J. pohliana—Medium shrub, purplish foliage, dense flowers.

FAMILY Lythraceae
GENUS *Lagerstroemia*
COMMON NAME Crepe myrtle

Appearance: Mainly trees, deciduous or evergreen with showy purple flowers.

Establishment: Easy to grow provided soil kept moist. Prefer well drained, friable, fertile soil but adapt to poorer soils. Propagate by semi-hardwood or hardwood cuttings over heat, or by seed.

Culture: Prune back to old wood after flowering, to encourage flowering for next year.

Cultivars

L. indica syn. *L. elegans*—Shrubby (1m) to small trees 3–6m tall. Variety of flower colours (eg. pinks, reds, purplish). Deciduous in cooler weather.

L. speciosa syn. *L. flos-reginae*—To 20m+. Suited to wet tropics, but will grow in sub-tropics. Flowers purplish or white.

FAMILY Verbenaceae
GENUS *Lantana*
COMMON NAME Shrub verbena

Appearance: Sprawling open shrubs, some prickly. Heads of colourful flowers, with scented foliage.

Note: Some varieties have become a serious weed problem in Queensland and New South Wales, self-seeding in bushland and other areas.

Establishment: Generally hardy and adaptable to most soils. Propagate by cuttings to ensure reliable flower colour.

Culture: Prune regularly to control spread.

Cultivars

Approx. 150 species, plus cultivars.

L. camara—To 2m tall. Typically yellow flowers, many named forms varying in growth habit and flower colour also exist.

L. montavidensis syn. *L. sellowiana*—A sprawling shrub to 1m+. Typically reddish purple flowers.

FAMILY Ampelidaceae
GENUS *Leea*
COMMON NAME Indian holly

Appearance: Shrubs with coloured decorative leaves.

Establishment: Needs good drainage, responds to fertile soil. Many species prefer protected sites, with some shade. Grows well in pots. Propagate by cuttings of spring side shoots.

Culture: Keep soil covered with thick mulch and water during dry periods.

Cultivars

L. amabilis—To 1m tall. Feathery. green-bronze leaves with white stripes on top and red below.

L. indica—Attractive, often multi-stemmed, evergreen, spreading shrub to 4m tall by 4m wide, with sparse, beautifully veined, pinnate leaves. Insignificant white flowers are followed by large clusters of dark-red to brownish black berries.

FAMILY Malpighiaceae
GENUS *Malpighia*

Appearance: Evergreen trees and shrubs with clusters of small white to reddish flowers and sometimes cherry-like fruits.

Establishment: Fertile well drained soil, full sun. Suitable as tub plants. Propagate by seed or semi-hardwood cuttings struck at 20–25°C.

Culture: Water during dry periods, prune for shape, feed annually.

Cultivars

Cultivated species include: *M. coccigera* (1m tall); *M. glabra* (to 4m); *M. nitida* (to 3m); *M. punicifolia* (to 3m) and *M. urens* (to 1.5m).

FAMILY Malvaceae
GENUS *Malviscus*
COMMON NAME Mallow

Appearance: Shrub, sometimes spreading, with reddish hibiscus-like flowers.

Establishment: Protect from strong wind, extreme cold or dryness. Full sun or light shade. Propagate by cuttings.

Culture: Responds to feeding, mulching, watering and pruning. Watch for insects.

Cultivars

Cultivated species include *M. arboreus* (to 4m) and *M. conzattii* (to 5m).

FAMILY Melastomaceae
GENUS *Medinilla*
COMMON NAME Kapa Kapa

Appearance: Evergreen shrubs or epiphytes. Decorative leaves have three prominent veins; flowers pink to reddish with yellow filaments.

Establishment: Prefer even temperature, light shade and a friable, fertile, moist soil. Propagate by seed or cuttings.

Culture: Like mulch, routine feeding. Prune after flowering.

Cultivars

M. curtsii—Grows to 1m.

M. magnifica—To 1m.

FAMILY Myrtaceae
GENUS *Melaleuca*
COMMON NAME Paperbark

Appearance: Evergreen trees and shrubs. Bottlebrush flowers less than 4–5cm diameter. Papery or flaking bark, stiff, pungent, scented foliage.

Establishment: Hardy and adaptable, with species suited to most climates. Respond to initial feeding, watering and mulching. Many tolerate cold, heat, wet and dry (within limits). Propagate mainly by seed.

Culture: Like mulch, respond to watering during extended drought. Prune lightly for shape. Few serious pests or diseases.

Cultivars

For wet tropics:

M. leucadendron syn. *M. leucadendra*—To 25m. Pendulous habit, white flowers.

M. linariifolia—To 10m. White bottlebrush flowers.

M. quinqenervia—To 20m+, cream flowers, attractive bark.

M. stypheloides—To 5m or taller, prickly leaves, white flowers.
M. viridiflora—To 15m, bright green flowers.

For dry tropics:

M. argentea—To 25m tall, pendulous foliage, cream flowers.
M. calothamnoides—To 2m, rounded shrub, red flowers.
M. cardiophylla—To 1m tall, cream flowers.
M. corrugata—To 1m, white flowers.
M. decussata—To 2m, unusual foliage, mauve flowers.
M. minutifolia—To 5m, small leaves, white flowers.

FAMILY Melastomaceae
GENUS *Melastoma*
COMMON NAME Lasiandra

Appearance: Shrubs and small trees. Showy purple, pink or white flowers, similar to *Tibouchina.*

Establishment: From humid tropic or sub-tropical rainforests. Will grow in full sun or light shade, provided soil is moist. Prefer friable organic soil. Protect from strong winds. Propagate by softwood cuttings at around 24–27°C.

Culture: Respond to mulch and watering in dry periods. Will take some pruning.

Cultivars

Approx 70 species.
M. candidium—To 3m. Fragrant, white to reddish flowers.
M. malabrathricum—To 2.5m, pink flowers.
M. sanguineum—To 6m, purple flowers.

FAMILY Rutaceae
GENUS *Melicope* syn. *Evodia*
COMMON NAME Evodia, Lacy lady

Appearance: Upright deciduous or evergreen shrubs, and some trees. Heads of small flowers.

Establishment: Prefer moist, fertile, well drained soil. Most are frost-tender. Full or filtered sun. Propagate by semi-hardwood, root cuttings, or by seed.

Culture: Susceptible to some leaf-chewing insects and scale, but rarely a serious problem for healthy plants.

Cultivars

20 species.
M. ternata—To 6m, greenish flowers.

FAMILY Rutaceae
GENUS *Murraya*
COMMON NAME Orange jessamine, Mock orange

Appearance: Trees and shrubs with large, scented, solitary flowers.

Establishment: Prefer friable, moist, well drained organic soil. Grown as hedge, tub plant or shrubbery. Full or filtered sun. Cold-hardy once established. Propagate by semi-hardwood cuttings at around 24°C.

Culture: Prune regularly to maintain shape. Mulch and water during dry periods.

Cultivars

M. paniculata—To 4m+ tall, with strongly scented white flowers and aromatic leaves.
M. koenigii (Curry leaf tree)—Scented flowers and leaves—leaves are used in curries.

FAMILY Rubiaceae
GENUS *Mussaenda*
COMMON NAME Bangkok rose

Appearance: Flowering evergreen shrubs with large showy flower bracts, usually shades of orange red or white.

Establishment: Does best in sub-tropics but will grow in protected, warm situations. Prefers friable organic soils, always moist but never saturated. Propagate by cuttings.

Culture: Water and mulch in dry weather, reduce watering in cooler weather. Lightly prune after flowering.

Cultivars

M. erythrophylla—To 40cm tall, yellow flowers.
M. macrophylla—To 2m, orange flowers.

FAMILY Apocynaceae
GENUS *Nerium*
COMMON NAME Oleander

Appearance: Dense bush with masses of brilliant flowers in many different colours. All parts of the plant are poisonous.

Establishment: Prefers friable well drained soil, but adaptable to most soils. Likes full sun. Propagate by semi-hardwood tip cuttings at 22–24°C.

Culture: Prune off previous year's growth after flowering.

Cultivars

N. oleander—To 6m tall, many named varieties.

N. indicum syn. *N. odorum*—To 2.5m, scented flowers

Family Ochnaceae
Genus *Ochna*
Common Name Carnival bush

Appearance: Evergreen trees and shrubs, with showy flowers and fruit.

Establishment: Moist, loose, organic soil. Like light to medium shade. Propagate by semi-hardwood cuttings.

Culture: Avoid compaction around roots (eg. from a parked car). Prune as needed. Prone to scale.

Cultivars

O. kirkii—To 5m, red flowers.

O. serrulata (Mickey Mouse plant)—Grows 1–2m. Yellow-green flowers with red calyx followed by red berries.

Family Oleaceae
Genus *Osmanthus*
Common Name Fragrant olive

Appearance: Evergreen flowering shrub, often fragrant.

Establishment: Best in a friable loam. Requires cool highland areas in the tropics. Will not grow well in hot, dry conditions. Propagate by cuttings from firm young growth, or by seed.

Culture: Needs lots of water while growing; reduce watering when growth slows. Prune only for shape.

Cultivars

O. fragrans—To 3m, fragrant flowers.

Family Acanthaceae
Genus *Pachystachys*
Common Name Cardinal's guard, Lollipop plant

Appearance: Open shrubs, flowers in varying colours.

Establishment: Sheltered, semi-shaded moist position. Propagate by semi-hardwood to hardwood cuttings.

Culture: Protect from wind and drought. Pinch out new growth to encourage bushiness.

Cultivars

P. coccinea—To 1.5m, scarlet flowers.

P. lutea—To 1m, yellow flowers.

Family Rubiaceae
Genus *Pentas*
Common Name Star cluster

Appearance: Flowering shrubs with soft, ribbed leaves.

Establishment: Semi-shaded, moist position. Loves heat. Propagate by softwood cuttings at around 25°C.

Culture: Regularly fertilise and water. Remove spent blooms. Subject to scale and aphids.

Cultivars

P. coccinea syn. *P. bussei*—To 50cm, red flowers.

P. carnea syn. *P. lanceolata* Egyptian pentas—To 50cm, pink flowers.

P. lanceolata—Herb with red and white striped flowers.

P. mussaenoides—To 5m tall.

P. parviflora—To 70cm tall, orange-red flowers.

Family Pittosporaceae
Genus *Pittosporum*
Common Name Australian native daphne

Appearance: Evergreen trees and shrubs with dense, attractive foliage of variable colour. Clusters of small flowers, sometimes scented, followed by decorative berries.

Establishment: Hardy, adapt to most soils. Most prefer temperate climates, but some suit sub-tropics or tropics. Prefer sun or filtered light. Good hedge or windbreak plants. Propagate by cuttings.

Culture: Respond well to feeding, watering and mulching. Avoid soil becoming bone dry; avoid waterlogging. Few serious pests; prone to scale and sooty mould.

Cultivars

P. rhombifolium (Queensland pittosporum)—To 20m+. Attractive orange fruits all year in the tropics.

P. undulatum—To 12m tall, fragrant flowers, orange fruits.

FAMILY Plumbaginaceae
GENUS *Plumbago*
COMMON NAME Leadwort, Plumbago

Appearance: Sprawling herbs and shrubs with clusters of colourful tubular flowers.

Establishment: Prefer moist friable soil, otherwise hardy to humid or dry air. Tolerate full sun or shade. Propagate by softwood cuttings of side shoots, with a heel, at 20–24°C, or by division.

Culture: Need pruning to contain growth and keep foliage dense. Mulch in dry climates. Few pests or diseases.

Cultivars

P. capensis (*P. auriculata*)—To 3m+ tall, blue to white flowers.

P. indica syn. *P. rosea*—To 70cm tall, reddish flowers.

P. zeylanica—To 50cm tall, white flowers.

FAMILY Apocynaceae
GENUS *Plumeria*
COMMON NAME Frangipani, Temple tree

Appearance: Deciduous trees and shrubs, milky sap. Simple undivided leaves, clusters of large, showy, fragrant flowers.

Establishment: Adapt to wet or dry tropics. Need well drained soil, prefer full sun. Respond to some moisture during warm weather. Frost-tender, but will tolerate colder periods once established. Propagate by cuttings in spring.

Culture: Keep roots drier during cool weather. Prune weak or straggly shoots after flowering.

Cultivars

Approx 8 species, including many varieties.

P. alba—To 12m. 30cm long leaves, flowers white with yellow centre.

P. obtusa—To 7m tall. Leaves 15cm long, white flowers with yellow centre.

P. pudica—To 3m tall, pale yellowish flowers.

P. rubra—To 8m tall. Broad oval leaves 60cm long, white or yellow flowers with reddish or yellowish centres. The most common species, with many different varieties varying in flower colour.

FAMILY Araliaceae
GENUS *Polyscias* syn. *Terminatia*
COMMON NAME Panax, Aralia

Appearance: Shrubs and trees with decorative scented foliage. Leaves are generally pinnate, glossy-green or variegated, dissected and irregularly toothed. Rarely flower in cultivation.

Establishment: Prefer warm humid location, fertile moist soil, and some shade. Propagate by hardwood cuttings.

Culture: Respond to feeding, watering, mulching and occasional pruning. Often used for hedging. Some make good large container specimens.

Cultivars

P. balfouriana (*Aralia balfouriana*)—To 8m tall in wild, with stringy, upright stems. Generally a large shrub in cultivation. Geranium-like green leaves to 10cm wide, with greyish speckles or blotches.

P. balfouriana 'Marginata'—has white margins on the leaves.

P. alfouriana 'Pennocki'—has larger leaves.

P. ficifolia—To 3m tall, with the branches usually purplish in colour; dissected leaves to 30cm long.

P. guilfoylei (*Aralia guilfoylei*)—To 7m tall, with few branches. Leaves pinnate to 40cm long, individual leaflets to 15cm long, and usually with white splotches and margins. Several cultivars available.

FAMILY Lythraceae
GENUS *Punica*
COMMON NAME Pomegranate

Appearance: Deciduous trees and shrubs, edible and ornamental fruits. Propagate by seed or cuttings.

Establishment: Prefer fertile, moist, well drained soil.

Culture: Feed annually, mulch in hot places, water during dry periods. Few serious pests.

Cultivars

P. granatum—To 5m.

P. granatum 'Nana'—Compact bush to 0.7m.

Family Rosaceae
Genus *Raphiolepis*
Common Name Indian hawthorn

Appearance: Hardy flowering shrubs. Shiny leathery leaves, clusters of white or pinkish flowers.
Establishment: Does well in tropics and sub-tropics. Prefers well drained, organic, moist soil, but is adaptable. Propagate by semi-hardwood cuttings or layering.
Culture: Prune for shape. Mulch and irrigate during prolonged dry periods.
Cultivars
Raphiolepis x *Delacourii*—To 2m (many varieties of this cross).
R. indica—To 1.6m.

Family Linaceae
Genus *Reinwardtia*
Common Name Yellow flax

Appearance: Evergreen shrub with bell-shaped yellow flowers.
Establishment: Does well in moist sub-tropics. Prefers temperatures above 13°C, and well drained, moist soils. Propagate by cuttings with heel at around 24°C.
Culture: Responds to mulching, feeding and watering during hot dry periods, over 18–23°C.
Cultivars
R. indica—To 1m, most commonly cultivated.

Family Ericaceae
Genus *Rhododendron*
Common Name Azalea, Rhododendron, Vireya

Appearance: Variable genus of trees and shrubs with many different groups, suited to different climates and conditions. Most are evergreen, leaves generally stiff. Flowers are trumpet-shaped, often in large, showy clusters.
Establishment: Prefer filtered sun or shade, fertile, friable organic soil; drainage and moisture are both essential. Protect from strong winds. Propagate by semi-hardwood cuttings, 24–27°C (27° for tropical species).
Culture: Respond well to feeding, mulching, watering in dry weather. Mites and root rots are a problem; other problems are rarely very serious.
Cultivars: Approx 800 species, those most commonly grown in tropics being 'Vireyas'.
R. vireyas—a subgenus of rhododendron; most are identified by variety names.
Azaleas—a subgenus of rhododendron; most are small shrubs identified by variety names.

Family Rubiaceae
Genus *Rondeletia*
Common Name Rondeletia

Appearance: Evergreen shrub, flowering pink to red.
Establishment: Prefers well drained, organic soil and some shade from severe sun. Propagate by semi-hardwood cuttings.
Culture: Keep moist while growing. Prune after flowering, to shape.
Cultivars
R. amoena—To 1m, pink flowers.
R. cordata—To 2m, pink to red flowers.
R. gratissima—To 1m, fragrant pink flowers.
R. odorata—To 1.8m, reddish to yellow fragrant flowers.

Family Acanthaceae
Genus *Ruellia*
Common Name Ruellia

Appearance: Shrubs and herbs of diverse appearance, usually with large, attractive flowers, reddish to pink or purple colours.
Establishment: Moist soil; prefer protected position with some shade. Avoid temperatures below 13°C. Propagate by cuttings, seed or division.
Culture: Respond to mulch and water during dry periods.
Cultivars
Commonly grown cultivars include:
R. baikeriei—to 1m.
R. formosa—to 70cm, scarlet tube-shaped flowers.
R. herbstii—to 1m.
R. macrantha—to 2m.
R. portellae—to 30cm.

Family Acanthaceae
Genus *Sanchezia*
Common Name Sanchezia

Appearance: Upright or climbing shrub; large leaves with heavy, showy veins and flower heads.

Establishment: Prefers friable, freely draining organic soil, protection from cold and wind, ample light (avoid heavy shade). Propagate by cuttings under mist or fog.
Culture: Mulch and water well in dry weather. Give medium to light prune after flowering. Water less when growth slows in cooler weather.
Cultivars
S. speciosa—To 1.8m tall, yellow and red flowers; green or variegated white or yellow leaves

FAMILY Araliaceae
GENUS *Schefflera* syn. *Brassaia*
COMMON NAME Umbrella tree, Star leaf
Appearance: Evergreen shrubs or trees, usually with large decorative palm-like leaves. Single or multiple trunks. Small flowers occur in large heads.
Establishment: Prefer freely draining moist soils, full or lightly filtered sunlight. Commonly grown as indoor plants in cooler areas, where they will stay much smaller in size. Propagate by seed, cuttings or marcotting.
Culture: Can suffer scale, mealy bug or sooty mould. Prune to make more dense.
Cultivars
S. actinophylla—To 13m, attractive foliage.
S. arboricola—To 4m, showy foliage.

FAMILY Caesalpiniaceae
GENUS *Senna*
COMMON NAME Winter senna
Note: Cassia has now been split into *Senna* and *Cassia*. *Senna* has flowers emerging from the sides of shoots (axillary), and has flattened pods; *Cassia* has flowers at tips (terminal), and cylindrical pods.
Appearance: Mainly trees with yellow flowers and pods that are flattened in cross section.
Establishment: Well drained, moist soil, sunny location. Propagate from seed soaked in boiling water.
Culture: Withstands light frost. Prune lightly to shape.
Cultivars
S. spectabilis—To 20m, yellow flowers.

FAMILY Solanaceae
GENUS *Solanum*
COMMON NAME Nightshade, Kangaroo apple
Appearance: Mainly shrubs and herbs with showy white, blue, purple or yellow flowers. Sometimes ornamental berry-like fruits. Note: Though some *Solanums* are edible (eg. capsicum and tomato), many other species contain toxins and should never be eaten by people or animals.
Establishment: Variable in requirements, according to species. Some come from wet rainforests, some from deserts. Generally hardy and tolerant of various soil types. Most like freely draining soils. Propagate by seed or cuttings.
Culture: Respond well to feeding, watering and mulching. Few serious pests.
Cultivars
Suitable for tropics:
S. erianthum—To 8m tall, white flower, yellow fruit.
S. havanense—To 1.8m, blue flower and fruit.
S. hispidum—To 2m, stout spines, white flower.
S. marginatum—To 2.5m, purple-veined white flower.

FAMILY Bignoniaceae
GENUS *Spathodea*
COMMON NAME African tulip tree
Appearance: Evergreen tree with large orange to reddish flowers.
Establishment: Adaptable, prefers well drained fertile soil and sunny open position. Propagate by seed or cuttings.
Culture: Prune if necessary, as required. Water during extended dry periods. Relatively pest-free.
Cultivars
S. campanulata—To 20m, smaller in sub-tropics. Leaves to 10cm long; large showy red flowers. Also a yellow flowering form.

FAMILY Musaceae
GENUS *Strelitzia*
COMMON NAME Bird of paradise
Appearance: Clump-forming plant with large banana-like leaves and unusual lance-like flowers.

Establishment: Does well in sub-tropics. Needs moist but well drained soil, full sun or light shade. Propagate by division or seeds.
Culture: Keep mulched and watered during dry periods. Feed annually.
Cultivars
S. reginae—Foliage grows to 1m, flowers orange with purple or red to 2m tall.
S. nicholai—Foliage to 3m+, pale mauve to bluish flowers.

FAMILY Solanaceae
GENUS *Streptosolen*
COMMON NAME Orange browallia
Appearance: Dense, evergreen shrub, with terminal bunches of orange flowers.
Establishment: Hardy in tropics, sub-tropics or mild temperate areas. Frost-tender. Needs moist, drained soil. Propagated easily from semi-hardwood cuttings.
Culture: Water in dry periods. Prune after flowering, to shape and promote growth.
Cultivars
S. jamesonii—Small evergreen to 2m tall, with profuse bright orange flowers.

FAMILY Acanthaceae
GENUS *Strobilanthes* syn. *Goldfussia*
COMMON NAME Goldfussia, Mexican petunia
Appearance: Shrubs and herbs with colourful tubular flowers, commonly clustered on spikes.
Establishment: Need high temperatures, humidity and moist soil. Prefer some shade. Propagate by cuttings.
Culture: Irrigate in dry periods. Maintain a thick mulch in any soil prone to drying out. Prune to contain plants.
Cultivars
S. dyeranus—Shrub with violet flowers.
S. isophyllus—To 1m tall, flowers blue, white, or pink.

FAMILY Myrtaceae
GENUS *Syzygium*
COMMON NAME Lilly Pilly
Appearance: Evergreen trees; glossy, green leaves, fluffy flowers in clusters mostly white or cream. Berry-like clusters of colourful fruits.
Establishment: Prefers moist, freely draining, fertile, mulched soil, full or filtered sun. Propagate by seed or cuttings.
Culture: Tolerates pruning — sometimes grown as a hedge. Responds well to feeding, watering and mulching. Susceptible to scale and sooty mould.
Cultivars
Approx 500 species, and an increasing range of named cultivars.
S. coolminianum—To 6m. White flowers, showy blue berries.
S. floribundum syn. *Waterhousea floribunda*—To 30m tall. Weeping habit, white flowers and berries.
S. luehmanii—To 10m tall, bronze young foliage, red berries.

FAMILY Bignoniaceae
GENUS *Tecoma*
COMMON NAME Trumpet bush
Appearance: Shrubs or small trees with orange or yellow clusters of trumpet-shaped flowers. May self-seed and become a weed in warm, humid climates.
Establishment: Adapt easily to most soils, humid or dry air, from mild-temperate to tropical climates. Prefer warm, freely draining, moist position, protected from extreme heat, cold or wind. Propagate by seed or hardwood cuttings.
Culture: Irrigate during drought. Prune hard if necessary. Transplant easily.
Cultivars
T. alata syn. *T. smithii*—Yellow flowers tinged with orange.
T. capensis syn. *Tecomaria capensis*—To 1.8m, reddish-orange flowers.
T. stans—To 6m, deciduous, bright yellow flowers.

FAMILY Combretaceae
GENUS *Terminalia*
COMMON NAME Tropical almond
Appearance: Trees with insignificant flowers; some have edible fruit, some have attractive foliage.
Establishment: Needs a tropical climate with distinct dry and wet seasons. Does well in poor soils. Propagate by seed.

Culture: Responds well to watering and feeding. Does not transplant. Few serious problems.

Cultivars

T. catappa—Deciduous, to 25m tall. Bright green foliage turns red before falling.

T. platyphylla—To 20m tall. Semi-deciduous in dry. Unpleasantly scented flowers.

FAMILY Araliaceae
GENUS *Tetrapanax*
COMMON NAME Rice paper plant

Appearance: Shrubs or small trees with large, palmate leaves, velvety-white when young becoming more rusty on mature plants.

Establishment: Moderately hardy. New shoots annually. Provide a protected position. Propagate by cuttings.

Culture: Frost-tender, adaptable to most soils. Will not tolerate wind. Susceptible to red spider mites and thrips.

Cultivars

One species, *T papyriferus*—To 2m+, large clusters of white flowers

FAMILY Apocynaceae
GENUS *Thevetia*
COMMON NAME Lucky nut

Appearance: Shrubs or trees with clusters of large funnel-shaped flowers; toxic milky sap.

Establishment: Prefer fertile, well drained soil and full sun. Propagate by seed or cuttings.

Culture: Can suffer from scale or aphis. Respond well to mulching. Avoid transplanting.

Cultivars

T. peruviana (Yellow oleander)—Grows 6–9m tall, fragrant yellow or orange flowers.

FAMILY Melastomaceae
GENUS *Tibouchina*
COMMON NAME Lasiandra, Glory bush

Appearance: Shrub with large, solitary or panicle flowers, small spiral-shaped fruit.

Establishment: Tolerates very light frost; best in a subtropical or tropical climate. Prefers freely draining, moist, organic soil, full or filtered sun. Protect from strong winds. Propagate by semi-hardwood cuttings.

Culture: Hardy and relatively pest-free once established. Appreciates mulch and irrigation during dry periods. Prune to keep foliage dense and healthy.

Cultivars

T. grandifolia syn. *T. urvilleana*—To 2m+. Large foliage, large purple flowers.

T. laxa—To 2m tall, semi-climbing, purple flowers.

FAMILY Lamiaceae
GENUS *Westringia*
COMMON NAME Native rosemary

Appearance: Hardy shrubs with masses of white to mauve flowers.

Establishment: Adapts to all but waterlogged soils. Prefers good drainage, full sun and good ventilation. Good coastal plant, excellent as a hedge. Propagate by cuttings.

Culture: Water during drought. Mulch to keep roots cool. Prune for shape and to keep dense.

Cultivars

W. fruticosa—Grows 1–2m tall, fast growing, bluish-green foliage. White flowering and variegated varieties.

3

Palms and Palm-like Plants

These plants are an integral component in imparting a tropical appearance to a garden. Most palms are truly tropical plants, preferring wet tropic or sub-tropic climates to do their best. Some palms and plants that have a palm-like appearance will grow under harsher conditions. These may be used in cooler climates, harsher inland areas (eg. deserts) or as indoor plants, to create similar effects when other varieties of palms become difficult to grow.

When you plant a new palm it usually takes a period of time to settle in to its new position before starting to put on significant growth. In warm climates most of the plants listed below should grow reasonably fast once they have settled in (6–18 months after planting).

SELF-CLEANING OR NOT

Self-cleaning palms: These drop old fronds naturally.

Not self-cleaning: The old fronds still cling to the trunk after dying. This is both unsightly and provides places for cockroaches and other insects to breed. Fronds need to be physically removed, making more work for the gardener.

SOLITARY OR CLUMP-FORMING

Most palms have either a solitary trunk (ie. one trunk to a plant) or develop multiple trunks from the base, to form a more bush-like clump.

PINNATE OR FAN-SHAPED FRONDS

Most palms have either fan-shaped fronds or pinnate (divided feather-like) fronds.

PROPAGATION

True palms (Arecaceae family) are propagated by seed. Seed needs to be collected close to the time it is about to drop. If you pick it too early, it is less likely to germinate. Palm seed can be spasmodic in germinating. Generally speaking, some may germinate within a month or two of planting, but other seed from the same batch may progressively germinate over many months or even years. Seed needs to be kept moist and warm to achieve this progressive germination. Waterlogged or cold seed will tend to rot.

FAMILY Arecaceae
GENUS *Archontophoenix*
COMMON NAMES Alexander and Bangalow palms

Appearance: Tall, graceful palms, slender trunks, self-cleaning.

Establishment: Prefers humid, frost-free areas, fertile organic soils.Grows in shade or sun. Propagate by seed.

Culture: Mulch and irrigate during dry periods.

Cultivars

A. alexandrae (Alexander palm)—Grows 10–20m tall; 3m long fronds with silvery undersurface.

A. cunninghamiana (Bangalow palm)—Grows 8–20m tall. Narrower diameter at base than Alexander palm, green undersurface to fronds.

FAMILY Arecaceae
GENUS *Areca*
COMMON NAME Golden cane, Betel nut

Appearance: Slender trunks can be multiple or solitary; large pinnate fronds.

Establishment: Most are sensitive to cold and drought. Some grow well in shade. Prefers wet

humid climate. Seeds germinate relatively fast.
Culture: Irrigate and mulch during dry weather. Protect from frost.
Cultivars
Approx 50 species.
A. cathecu (Betel nut palm)—To 25m or taller, solitary trunk. Kernels are chewed as betel nut.
A. ipot—To 5m tall, solitary trunk.
A. lutescens syn. *Chrysalidocarpus lutescens* (Golden cane)—To 4m tall. Clump-forming dense, bushy appearance, attractive, yellowish, palmate fronds to over 2m long.
A. triandra—To 5m tall, multiple trunks.
A. vestiaria—To 9m tall, solitary trunk with stilt roots.

FAMILY Arecaceae
GENUS *Arecastrum* syn. *Syragus, Cocos plumosa*
COMMON NAME Queen palm
Appearance: Grey-brown solitary trunk to 20m tall; arching/drooping leaves.
Establishment: Grows well in wet sub-tropics or tropics. Prefers moist well drained soil, full sun. Propagate by seed.
Culture: Mulch and water in dry periods.
Cultivars
Only one species, *A. romanzoffianum*.

FAMILY Arecaceae
GENUS *Arenga*
Appearance: Dwarf or large solitary trunks, or forming clumps. Pinnate fronds, messy trunks covered with fibre. Pulp from fruits can irritate the skin.
Establishment: Native of tropical Asia, prefers hot wet climates. Propagates from seed; will also grow, with difficulty, from suckers.
Culture: Mulch and water in dry periods. Responds to annual feeding.
Cultivars
Approx. 17 species.
A. engleri—To 3m tall, bushy appearance.
A. pinnata (Sugar palm)—To 10m tall. A source of starch, sugar and alcohol in some areas.
A. undulatifolia—To 9m tall, clump-forming and spreading.

FAMILY Agavaceae
GENUS *Beaucarnea* syn. *Nolina*
COMMON NAME Ponytail palm or Bear grass
Appearance: Stiff strap-like leaves, as a crown on a trunk swollen at the base; overall sometimes resembling a palm.
Establishment: Grows well in warm drier climates; some may grow in wetter areas also. Propagate by seed.
Culture: Ensure good drainage and ventilation. Water freely when growing but much less when growth slows.
Cultivars
B. recurvata syn. *Nolinia recurvata* (Ponytail palm)—To 10m tall, leaves nearly flat.
B. stricta—To 5m tall, leaves V-shaped in cross section.

FAMILY Arecaceae
GENUS *Bismarckia*
COMMON NAME Bismarck palm
Appearance: To 8m or taller, and 1.2m or longer blue-green fan-shaped fronds.
Establishment: Generally easy to grow in frost-free tropical or subtropical wet climates. Prefers good drainage. Avoid transplanting larger plants.
Culture: Seed usually takes up to 3 months to germinate.
Cultivars
Only one species; *B. nobilis* syn. *Medemia nobilis*.

FAMILY Arecaceae
GENUS *Carpentaria*
COMMON NAME Carpentaria palm
Appearance: Smooth, ringed trunk 14 to 20m tall. Elegant arching, pinnate fronds.
Establishment: Only wet tropical climates. Preferably moist, organic-rich soil.
Culture: Needs lots of water, does not tolerate cold or drought. Will grow in coastal areas and has been used as a street tree with some success.
Cultivars
Only one species, *C. acuminata*, native to northern Australia.

FAMILY Arecaceae
GENUS *Caryota*
COMMON NAME Fishtail palm

Appearance: Leaves divided into triangular leaflets. Can form multi-trunked clumps or have solitary trunks. Has many ornamental uses due to its unusual bipinnate foliage.

Establishment: Most species grow better in tropical or warmer areas (temperatures always above 5–7°C, even at night). Most seeds germinate in 3 months.

Culture: Grows well on hillsides or sloping areas, and often quite quickly. Species is hardy from tropical to warm temperate climates, but heavy or frequent frosts will affect them adversely.

Cultivars

C. cumingii—To 8m tall; dense, clump-forming habit.

C. mitis—To 12m tall, thin trunk. Will tolerate shade. Can die after producing seed.

C. no—To 25m tall, thin trunk.

C. ochlandra—Relatively cold-tolerant, can be grown in Sydney.

FAMILY Arecaceae
GENUS *Chamaedorea*
COMMON NAME Bamboo palm, Parlour palm

Appearance: Tidy green bamboo-like trunks. Varies in size, foliage and habit. Can be solitary or clump-forming, generally pinnate fronds (some exceptions).

Establishment: Prefers humid, shaded and protected area. Seed can germinate within 6 months of sowing if sown fresh.

Culture: Tolerates low light and dry periods once established. Grows well as indoor or understorey plant. Will tolerate poor treatment, but will respond well to small amounts of fertiliser. Fruits contain a skin irritant. Caterpillars and mites may become a problem if grown outdoors.

Cultivars

Approx. 100 species.

C. elegans (Parlour palm)—To around 2m tall, thin bamboo-like trunks. Likes shaded, warm and moist conditions.

C. erumpens (Bamboo palm)—To 3m tall. narrow dark green trunks, clump-forming; feather frond.

C. geonomaeformis—Solitary slim dark green trunk to 1.2m. Frond simple—divided into halves only. Slow-growing.

C. seifrizii (Bamboo palm)—Clump-forming to 3m tall. More sun tolerant than other species.

FAMILY Arecaceae
GENUS *Chamaerops*
COMMON NAME Mediterranean fan palm

Appearance: Solitary trunk, to 6m tall, or low multi-trunked clumps. Stiff dark green fan-shaped fronds.

Establishment: Very hardy; adapts to most soil types, tolerates extreme heat and even snow. Prefers full sun. Does especially well in free-draining soils in a mild climate. Seed germinates in 3–4 months.

Culture: Requires little attention once established.

Cultivars

One species, *C. humilis*, native to Europe.

FAMILY Arecaceae
GENUS *Chambeyronia*

Appearance: Tall, solitary palm from New Caledonia. Distinctive bright red leaves when newly formed give striking ornamental effect.

Establishment: Requires protection from wind when young and moderate to high watering. Mulching is recommended.

Culture: Slow-growing genus that is adaptable to wide range of climates from warm temperate to tropical.

Cultivars Only 2 species, the most commonly grown being *C. macrocarpa*, to 18m tall. Its large leaves are reddish when young, becoming green with age.

FAMILY Arecaceae
GENUS *Cocos*
COMMON NAME Coconut palm

Appearance: Single, often bent, yellowish, self-cleaning trunk, topped with large stiff fronds. Various shapes, sizes, and leaf colours are available.

Establishment: Best on coastal foreshores from warmer sub-tropics to tropics. (Grows on the

Gold Coast, but needs warmer humid climate to produce coconuts.) High fatality rate amongst young plants in cooler climates. Plant in spring for best results. Propagate by planting seed (coconuts) in permanent position.
Culture: Water and mulch during drought.
Cultivars
Only one species, *Cocos nucifera*, but both tall (10m) and dwarf (4–5m) forms exist.

FAMILY Agavaceae
GENUS *Cordyline*
COMMON NAME Grass tree
Appearance: Single or multi-branched trunk topped with strap-like leaves. Some species retain leaves over much of the trunk. Bush or tree-like, according to species.
Establishment: Generally adapts to a wide variety of climates from temperate to tropical areas. Most tolerate a wide variety of soils, and sun or shade.
Culture: Responds to feeding and, during dry periods, watering and mulching. Propagate by seed or cuttings.
Cultivars
A wide range of species and named varieties occur, and more are being introduced all the time. Many of these have attractive coloured foliage which tends to increase in intensity during cooler weather.

FAMILY Arecaceae
GENUS *Corypha*
COMMON NAME Talipot palm, Gebang palm
Appearance: Tall, robust, solitary trunk marked with attractive leaf scars. This genus is the largest of the fan palms.
Establishment: Only grows in wet monsoonal tropics. Does well on flat, low-lying ground. Seed germinates from 2 to 3 months.
Culture: Needs replacing periodically, as it dies after flowering, although this may take 20 years or more to occur.
Cultivars
C. elata syn. *C. utan* (Gebang palm) — To 22m.
C. umbraculifera (Talipot palm) — To 18m.

FAMILY Cycadaceae
GENUS *Cycas*
COMMON NAME Cycads
Appearance: Glossy palm-like fronds emerge from a central trunk. Seed borne on single cone which emerges in centre of the fronds, atop trunk.
Establishment: Best in deep, fertile, well drained soils. Can be slow growing, but very long lived. Propagated by slow-germinating seed.
Culture: Mulch and water during dry periods.
Cultivars
C. circinalis (Sago palm)—To 2.5m tall.
C. media (Australian nut palm)—To 6m tall.
C. normanbyana—To over 3m tall.
C. revoluta (Japanese sago palm)—To 3m tall.
C. rumphii—To 6m tall.

FAMILY Arecaceae
GENUS *Cyrtostachys*
COMMON NAME Sealing wax palm
Appearance: Slender trunks, clump-forming habit. Some species very ornamental due to elegance and colour of stems.
Establishment: Prefers moist, even swampy soil, and some shade. Only suited to tropics. Propagate seeds in hot bed (germinates in 2–3 months) or by dividing suckers and growing on in warm sheltered position or glasshouse environment.
Culture: Needs protection from cold if temperatures drop below 15°C.
Cultivars
C. lakka—To 4m or taller, fronds around 1.5m long with greyish undersurface.
C. renda (Lipstick palm)—To 10m tall, reddish top to otherwise greenish brown stems.

FAMILY Zamiaceae
GENUS *Dioon*
COMMON NAME Cycad
Appearance: Typical palm-like habit with stiff, pinnate leaves and woolly cones.
Establishment: Prefers partial shade and fertile, friable soil. Slow-growing, needs constant moisture while establishing. Propagate by seed.
Culture: Mulch and water during dry periods.

Cultivars

D. edule (Chestnut dioon)—To 2m tall; leaves to 2m long, edible seeds.

D. purpusii—To 1m tall, spreading foliage.

D. spinosum—To 7m or taller; relatively slender trunk, spiny foliage.

FAMILY Agavaceae
GENUS *Dracaena*
COMMON NAME Dragon tree

Appearance: Shrubby or tree-like, with strap-like leaves and narrow trunks.

Establishment: As for *Cordylines*.

Culture: As for *Cordylines*.

Cultivars

There are many cultivars, with varying foliage colour and texture, grown widely in the tropics and sub-tropics. They are closely related to *Cordylines*, only distinguished by a slight difference in the flowers and fruits.

FAMILY Arecaceae
GENUS *Dypsis* syn.. *Chrysalidocarpus*

Appearance: Small, thin palm with small, reddish-green leaves.

Establishment: Shade-loving plant that also requires adequate warmth and moisture in order to thrive. Propagate from seed; takes 2 months or longer to germinate. Bottom heat may be of benefit.

Culture: Mulch and water in dry weather. If growing in containers, move into more protected position if weather becomes cooler.

Cultivars

D. pinnatifrons syn. *D. gracilis* — To 5m, very dainty green to red-green leaves.

FAMILY Arecaceae
GENUS *Howea*
COMMON NAME Kentia or Sentry palm

Appearance: A feather palm with a long, slender trunk and highly ornamental and draping foliage.

Establishment: Needs part shade while young; will grow permanently in some shade. Requires constant moisture, but also well draining soil. Does well in coastal areas. Fresh seed germinates in 2–12 months.

Culture: Does not tolerate drought, frosts or hot, dry winds.

Cultivars

Only two species.

H. belmoreana (Belmore sentry palm)—To 6m tall.

H. forsteriana (Kentia palm)—To 15m tall. Very popular indoor plant.

FAMILY Arecaceae
GENUS *Hyphorbe* syn. *Mascarena*
COMMON NAME Bottle palm, Spindle palm

Appearance: Stiff, recurved, feather-leaved fronds. Base of trunk swollen. Not self-cleaning.

Establishment: Hardy in tropical climates, needs full sun and cold-free area.
Propagates easily from fresh seed.

Culture: Can be grown as an indoor plant if provided with ample light, warmth and moisture. Tolerates salt winds and coastal conditions.

Cultivars

H. lagenicaulis (Bottle palm)—To 3m tall, thick, bulging trunk.

H. verschaffeltii (Spindle palm)—To 6m tall.

FAMILY Arecaceae
GENUS *Licuala*

Appearance: Attractive circular or rectangular fan-shaped leaves, sometimes divided, sometimes not. Can have solitary or multiple trunks.

Establishment: Needs wet, warm climate, preferably with some shade particularly while young. Seed can take 3–6 months to germinate; some species by division of clusters.

Culture: Does not tolerate cold, dry, hot sun or windy conditions.

Cultivars

L. gracilis—To 1.2m tall, clump-forming.

L. grandis (Ruffled fan palm)—To 3m tall, solitary trunk.

L. ramsayi syn. *L. meulleri*—Grows 5–12m tall, solitary trunk.

FAMILY Arecaceae
GENUS *Livistona*
COMMON NAME Fan palm

Appearance: Large fan-shaped fronds, solitary trunk.

Establishment: Some relatively hardy species will even tolerate frosts. Prefers moist soil such as found along streams, swamps or estuaries. Seed can germinate within 6 weeks.
Culture: Most species enjoy full sun conditions although shade-tolerant species that are natives of forest fringes do exist.
Cultivars
L. australis (Cabbage tree palm)—To 20m tall. Very hardy species, tolerates frost.
L. chinensis (Chinese fan palm)—To 18m tall, broad trunk.

Family Zamiaceae
Genus *Macrozamia*
Common Name Cycad
Appearance: Palm-like, trunk rarely branched. Spine-tipped foliage distinguishes these from many other cycads.
Establishment: Many species occur naturally in arid or semi-arid areas; a few come from less arid areas. Most will grow in full or filtered sun. Generally avoid heavy shade. Adapt to most soils, but prefer reasonable drainage. Seeds form readily and germinate easily, but often very slowly.
Culture: Mulch and water during dry periods. Slow to establish but, provided they are not waterlogged, they grow with little attention.
Cultivars
M. communis—To 2m tall, suits humid subtropics.
M. lucida—To 1.2m. Occurs naturally in south-east Queensland.
M. macdonnellii—To 3m tall, from arid central Australia.

Family Arecaceae
Genus *Neodypsis*
Common Name Triangle palm
Appearance: Beautiful pinnate fronds, similar in appearance to *Chrysalidocarpus*. Fronds attached at a sharp angle to the trunk.
Establishment: Prefers warmer humid subtropics or tropics; does not tolerate cold very well. Can be grown in protected sites in south-east Queensland. Germination of seed occurs within 1 month.
Culture: Keep moist and mulched in dry weather; protect from cold winds.
Cultivars
N. decaryi (Triangle palm)— Grows 5–10m tall. The most common species will do well on protected coastal sites in south-east Queensland, but better in the tropics. Has grey-green fronds and unusual trunk base which is triangular in cross section, making this a highly sought after palm.
N. lastelliana—To 7m or taller, with reddish colour on leaf stalks and crown shaft. Hardier than *N. dacaryi*.

Family Arecaceae
Genus *Normanbya*
Common Name Queensland black palm
Appearance: Plume-like, dense crown atop a long, slender trunk.
Establishment: Best if established in a sheltered site with some degree of shading, from which it will grow through into a full sun position with maturity.
Culture: Seeds can deteriorate rapidly, so they must be sown fresh. Young trees have unusual, somewhat plain, broad wedge-shaped leaflets, making the plant something of an 'ugly duckling', as it develops into a very different and far more attractive tree at maturity.
Cultivars
N. normanbyi (Queensland black palm)—Solitary, smooth trunk to 20m tall, plume-like feathery fronds. Prefers humid tropics.

Family Pandanaceae
Genus *Pandanus*
Common Name Screw pine
Appearance: Palm-like or clump-forming, with long or short trunks, topped with a cluster of strap-like leaves.
Establishment: Generally tropical or subtropical coastal areas. Adapts to most soils, provided they are well drained. Salt-tolerant. Propagate by seed soaked in water for 24 hours prior to planting, or by removing suckers.
Culture: Once established, relatively hardy. Water in extended dry periods. Borers have

devastated *Pandanus* in some places (eg. south-east Queensland).

Cultivars

Over 650 different species.

FAMILY Arecaceae
GENUS *Phoenix*
COMMON NAME Date palm

Appearance: Size can vary from a large bush to very tall specimens. Usually solitary trunk; sometimes pinnate fronds have spines.

Establishment: Some of the hardiest palms to cold and dry, many come from arid areas. Grow well in containers (inside or out). Best in a moist, drained soil.

Culture: Not self-cleaning, so old fronds need to be cut off as they die—a major task on tall specimens.

Cultivars

P. canariensis (Canary Island date palm)— Thick solitary trunk to 20m tall. Tolerates cold (eg. southern Australian states). Green fronds to 6m long.

P. dactylifera (Commercial date)—Thick trunk, to 20m tall. Sparse head of grey-green fronds. Can sucker from base. Needs hot, arid conditions to crop.

P. reclinata (African wild date)—Clump-forming, 7m tall, 3m long fronds.

P. roebelenii—To 3m tall, with 1.5m long fine-textured fronds with spikes.

P. rupicola (Cliff date palm)—To 5m tall, leaves to 3m long. A particularly graceful species.

FAMILY Arecaceae
GENUS *Pritchardia*
COMMON NAME Fan palm, Lou, Lou-lu, Loulou Lelo palm

Appearance: Straight, clean, tall or medium-sized solitary trunks. Almost circular fan-shaped fronds with delicate frayed outer edge, beautiful texture and landscaping potential. Closely related and similar to *Washingtonia*.

Establishment: Requires a moist tropical or protected subtropical position. Loves the sun, does not tolerate cold. Some hardier species will grow as far south as Sydney in warm, protected positions.

Culture: Seed takes 2–3 months to germinate under ideal conditions. An Australian native form grows in high rainfall, on sloping hills and valleys with volcanic soils.

Cultivars

P. pacifica (Fiji fan palm)—To 10m tall, thin trunk, 1m diameter fronds.

P. hillebrandii, (Loulou Lelo palm)—A hardier species to 7m in height.

P. martii—To 5m, a hardier species.

FAMILY Arecaceae
GENUS *Ptychosperma*
COMMON NAME Solitaire palm, Macarthurii palm, Hurricane palm

Appearance: Pinnate fronds, solitary trunks or clump-forming. Distinctive crown shafts, bunches of colourful red fruits.

Establishment: All are tropical plants, so establishment may be difficult even in the sub-tropics. Need warmth, moisture and protection from drying wind. Good in shade. Will do well in containers or the ground, provided the air is humid. Seed germinates within 3 months. Clump-forming species can be propagated by division.

Culture: Water and mulch in dry weather. If grown in pots, move into protected, warm position during cold weather.

Cultivars

P. elegans (Solitaire palm)—Solitary grey trunk to 12m tall.

P. macarthurii (Hurricane palm)—Very slender trunk to 7 m.

FAMILY Arecaceae
GENUS *Rhapis*
COMMON NAME Lady palm

Appearance: Multiple, slender cane like trunks forming clumps; palmate, lobed leaves. Generally a smaller palm used indoors, in small gardens or as an understorey plant. Variegated species are available but are quite rare and need to be propagated by division.

Establishment: Grow best in light shade (but can tolerate direct sun). Need moist but not over-wet soil. Often grown as container plants for inside or out. Seed germinates

within 3 months; alternatively, divide clumps.
Culture: Avoid overwatering, particularly in cooler weather. Keep moist but never flooded. May benefit from light application of lime if in a very acidic soil (eg. if you apply manure or compost, add a little lime to counteract the acidity). Can suffer from scale or mealy bug.

Cultivars

R. excelsa (Lady palm)—To 3m tall, multiple stems; deeply, lobed fan-shaped fronds.

FAMILY Strelitziaceae
GENUS *Ravenala*
COMMON NAME Traveller's palm

Appearance: A clump of banana-like leaves atop large palm-like trunks.
Establishment: Grows easily in wet tropics and sub-tropics.
Culture: Generally easy once established. Remove dead leaves occasionally, and water in dry weather.

Cultivars

One species, *R. madagascariensis*, to 10m tall, leaves to 3m long.

FAMILY Arecaceae
GENUS *Roystonia* syn. *Oreodoxa*
COMMON NAME Royal palm

Appearance: Solitary trunks that are often grey and bulging to a certain level and then followed by a slender green shaft to the crown. Dense crown of long arching fronds.
Establishment: Prefers warm, wet, tropical climates. Grows more slowly in sub-tropics. Seed can germinate within 2 months.
Culture: Mulch and water in dry periods. Apply well rotted manure or slow-release fertiliser regularly.

Cultivars

R. oleracea (Carribean royal palm)—Very tall palm, up to 40m.
R. regia (Cuban royal palm)—To 25m tall. Distinctive, clean, shiny green shaft at the top the trunk, above grey slightly bulging lower trunk; fronds up to 4m long.

FAMILY Arecaceae
GENUS *Veitchia*
COMMON NAME Manilla palm

Appearance: Mostly very tall and slender feather palms with drooping pinnate leaves; solitary trunks with distinctive rings. Overall very clean and attractive plants.
Establishment: From tropical rainforests. Need protection from wind, plenty of water and are not very tolerant of cold winds or frosts. Fresh seed germinates in 1–2 months.
Culture: Useful, fast-growing palm that has a number of landscaping possibilities.

Cultivars

V. arecina—To 12m tall.
V. joannis—To 30m tall, arched leaves to 3m long.
V. merrillii—To 3m tall, fronds to 2m long.
V. montgomeryana—To 12m or taller, fronds to 3m long.

FAMILY Arecaceae
GENUS *Verschafeltia*
COMMON NAME Stilt palm

Appearance: A dark and slender trunk that has exposed stilt-like roots at its base. This feather palm is quite eye-catching in its native sloped environment, often protected by surrounding rainforest.
Establishment: Will survive in rocky soils but does require some protection from cold winds and frosts. Requires ample watering at all stages of growth.
Culture: Seed will germinate between 2 and 3 months under warm and humid conditions.

Cultivars

One species, *V. splendida*, to 25m tall. Attractive but spiny black trunk which develops stilt roots. Fronds have spines on the sheath. Grows well in Darwin.

FAMILY Arecaceae
GENUS *Washingtonia*
COMMON NAME Cotton palm

Appearance: Sturdy, thick-set, solitary trunk; fan-shaped fronds forming large, clumped crown.
Establishment: Tolerates both heat and cold. Prefers full sun, well drained soil and an open position. Grows in humid or dry environments, but needs some soil moisture. Will grow in tubs,

but these are deep-rooted palms, so minimise transplanting. Propagate by seed.

Culture: Needs some moisture in the ground during dry periods, otherwise hardy and requires minimal attention.

Cultivars

W. filifera—To 15m, thick grey trunk. Tolerates drought, heat and frost.

W. robusta—To 35m tall, brown trunk. Fronds have thorns. Not as hardy as *W. filifera*.

FAMILY Arecaceae
GENUS *Wodyetia*
COMMON NAME Foxtail palm

Appearance: Single-trunked feather palm with droopy, brushy type fronds that resemble a fox tail and give this palm its common name.

Establishment: Prefers full sun and tropical climate but will tolerate warm temperate climate without frosts. Likes well drained soil, though young specimens require a lot of water. Older plants can better withstand harsh climatic conditions. Propagation by seed should be undertaken in warm area, with germination after 2 months.

Culture: Relatively new in cultivation, so information is still uncertain. Deep, infrequent waterings probably needed in extended drought.

Cultivars

Only one species; *W. bifurcata* (Foxtail palm) grows to 15m.

FAMILY Agavaceae
GENUS *Yucca*
COMMON NAME Dagger plant, Needle palm

Appearance: Clump of stiff, strap-like leaves atop stem, often sharp points on leaves; sometimes branching woody trunks. Large spikes of white or purplish flowers on mature plants.

Establishment: Requires good drainage. Prefers sandy loam, open ventilated position, does well in dry, hot places. Some species will tolerate more humidity or more cold than others. Propagate by seeds, cuttings, rhizomes or root division.

Culture: Flowers best in hot, dry conditions. In wetter climates, it may do better in stone-mulched, well drained raised bed.

Cultivars

Y. aloifolia—Single or branched trunk to 7m tall. Several named forms varying in foliage colour. Does well in dry sub-tropics.

Y. filamentosa—To 5m tall, suits humid or dry sub-tropics.

FAMILY Zamiaceae
GENUS *Zamia*
COMMON NAME Cycad

Appearance: Mainly low palm-like plants from tropical and subtropical America. Foliage varies in appearance according to age and size of plant, sometimes strap-like, or broad.

Establishment: Usually an understorey plant so shady site that gives protection from strong winds (hot, dry or cold) is required.

Culture: Many species thrive in alkaline soils. Propagation from seed (nuts or cones) in moist propagating soil, with shoots being repotted individually.

Cultivars

Approx. 40 species, difficult to distinguish apart without cones.

4

Bamboos, Grasses and Grass-like Plants

These plants are generally very hardy, particularly once established. They can add a different mood to a garden through the typically soft, narrow foliage, contrasting with the broader leaves of most other tropical garden plants. Many species can be invasive, blocking pipes, lifting paving or cracking walls. Bamboos are particularly notorious. Problem plants like this should be grown away from such structures, or kept contained to a walled planter box or container. While the vigour of bamboos can be a threat, it also makes them fast-growing and relatively easy to care for. Often a bamboo will survive as a tub plant where all else fails.

FAMILY Poaceae
GENUS *Arundinaria*
COMMON NAME Bamboo
(Note this is one of several genera commonly called bamboo).
Appearance: Hardy, clump-forming bamboos with persistent culm or stem sheaths. In some, young shoots are colourful.
Establishment: Most species prefer part or full shade. Avoid very heavy or hard soils; best in friable soils of average or good fertility. Avoid extreme cold, wind or dry conditions. Better suited to sub-tropics and temperate climates than some other bamboos. Propagate by division.
Culture: Need little attention apart from thinning out to control spread.
Cultivars
A. falcata syn. *Chimonobambusa falcata*—To 6m tall, yellowish stems.
A. gigantea—To 8m tall. Leaves have fine-toothed margins and are about 30cm long.
A. vagans syn. *Bambusa pygmea* — To 40cm tall. Can be invasive.

FAMILY Poaceae
GENUS *Bambusa*
COMMON NAME Bamboo
Appearance: Typical bamboo clump, usually with a branching habit, culms (stems) are commonly, but not always, green. Various sizes, up to 15m tall. In cooler areas, culms may be produced later in summer.
Establishment: Best suited to tropics and sub-tropics but some are hardy into temperate zones. Likes full or filtered sun. Propagate by division.
Culture: Needs little attention apart from thinning out to control spread.
Cultivars
Over 100 species, and many varieties available.

FAMILY Cyperaceae
GENUS *Carex*
COMMON NAME Sedge or Blue grass
Appearance: Hardy herbaceous grass-like, clump-forming sedge. (Note: It is illegal to import any non-native *Carex* into Australia).
Establishment: Grows well in most soils, but prefers wet conditions (eg. edge of a pond). Propagate by seed in spring, or division any time.
Culture: Grows easily in temperate climates with little threat of becoming a weed, however may naturalise in warmer climates. Responds well to mulching and watering in dry periods. Cut out dead foliage and flower heads.
Cultivars
Many species are in cultivation in temperate

Parc Floral de Nice

2. *Codiaeum* Croton

Mediterranean style

4. *Murraya paniculata* displaying trunk

5. *Allocasuarina verticillata* (syn. *Casuarina stricta)*

6. *Ardisia crenata*

7. *Ardisia crispa* (syn. *A. crenata)* Spear flower

8. *Calliandra haematocephala* Powder puff tree

9. *Camellia sasanqua* 'Yuletide'

Cestrum nocturnum Night-scented jessamine

11. *Grewia caffra* Lavender star

Gardenia jasminoides Gardenia

Graptophyllum pictum Caricature plant

14. *Grevillea* 'Sandra Gordon'

15. *Ixora* sp. Flame flower

16. *Plumeria rubra*

17. *Pachystachys lutea* Cardinal's guard

18. *Rhododendron vireya* 'Sambu Sunset'

19. *Zamia furfuracia*

. *Dracaena* cv. Dragon tree

21. *Hyophorbe lagenicaulis* Bottle palm

. *Arecastrum romanzoffianum* Queen palm

23. *Washingtonia robusta* Cotton palm

24. *Cocos nucifera* Coconut palm

25. *Cordyline* 'Hawthorn Rose'

26. *Macrozamia* sp. Cycad

27. *Wodyetia bifurcata* Foxtail palm

28. *Cocos nucifera* Dwarf coconut

. *Chamaedorea seifrizii*

30. *Yucca aloifolia* Dagger plant

. *Normanbya normanbyi* Black palm

32. *Phoenix canariensis* Canary Island date palm

33. *Bismarckia nobilis* Bismarck palm

34. *Phyllostachys aureosulata* 'Spectabilis'

35. *Bambusa balcooa* Bamboo

. *Phyllostachys aurea* Golden bamboo

37. *Taxodium distichum*

. *Araucaria heterophylla* Norfolk Island pine

39. *Thuja orientalis aurea* Thuya

40. *Juniperus chinensis pyramidalis* Juniper

41. *Juniperus conferta* 'Emerald Sea' Prostrate juniper

42. *Podocarpus elatus* Yellow wood

43. *Averrhoa carambola* Star fruit

44. *Psidium guava* Guava

. *Ananas comosus* Pineapple

46. *Musa* spp.

7. *Citrus paradisi*

48. *Diospyros kaki* Oriental persimmon

49. *Diospyros digyna* Black sapote

. *Capsicum annuum*

51. *Solanum seaforthianum*

52. *Thunbergia alata* Black-eyed Susan

3. *Ficus pumila* Fig

54. *Ipomoea horsfalliae* Cardinal creeper

55. *Bougainvillea* cv.

56. *Pyrostegia venusta* Golden shower

57. *Rhoeo spathacea* Purple-leafed spider wort

58. *Russelia equisetiformis* Cigarette plant

59. *Aglaonema* sp. Chinese evergreen

60. *Anthurium* cv. Flamingo plant

1. *Selaginella martensii*

2. *Crinum procerum* var. *Kaawanum*

63. *Zephranthes candida* Zephyr lily

64. *Sprekelia formosissima* Jacobean lily

65. *Dichorisandra thyrsiflora* Blue ginger

66. *Tacca chantrieri* Bat flower (white form)

67. *Alpinia purpurata* Red ginger

68. *Kniphofia* cv. Torch lily

climates; some may be suited to tropics and sub-tropics.

C. acuta—To 70cm tall.

C. baccans—To 1m tall, attractive purplish fruits.

C. pseudo-cyperus—To 90cm tall.

C. tristachya—To 25cm tall, white variegated leaves.

FAMILY Poaceae
GENUS *Cortaderia*
COMMON NAME Pampas grass

Appearance: Large, hardy clump-forming perennial grasses, with sharp-edged leaves.

Establishment: Many will grow in most soils, but prefer moist, fertile sandy soils. Propagate by seed.

Culture: Responds well to watering at dry times. Decorative plumes are harvested and used for dried flower arrangements. Some species have become weeds in some areas.

Cultivars

C. rudiuscula—To 1.8m.

C. selloana—To 5m tall, white or pinkish plumes.

FAMILY Liliaceae
GENUS *Dianella*
COMMON NAME Flax lily

Appearance: Fibrous-rooted clump-forming perennial; flax-like narrow leaves, branched spikes of bluish flowers with yellow anthers to 1m tall, Small round fruits, often bright blue.

Establishment: Generally hardy, and easily established in sun or shade. Some species are more suited to the sub-tropics and tropics than others. Propagate by division or seeds.

Culture: Requires little attention once established.

Cultivars

D. caerulea—Suits tropics, occurs naturally in shade of rainforests.

D. ensifolia (*D. nemorosa*)—Grows 1–1.8m tall, blue or white flowers.

FAMILY Poaceae
GENUS *Imperata* syn. *Miscanthus*
COMMON NAME Japanese blood grass

Appearance: *Imperata* is the more popularly known name, though in fact it is not botanically correct!. The genus *Miscanthus* includes a number of ornamental grasses commonly grown for their attractive silky heads. Leaf blades are long, flat, and can vary considerably in height according to species.

Establishment: Deciduous in temperate climates, dying back over winter. Though not widely grown in warmer areas, it has potential in the sub-tropics and perhaps the tropics. Grows well in pots, if kept moist. Likes full or lightly filtered sun light. Propagate by division.

Culture: Responds well to mulching, feeding and watering. Will spread by underground roots, but is probably slow to become a serious problem.

Cultivars

The most promising cultivar in Australia is *Imperata cylindrica* 'Rubra' syn. *Miscanthus sacchariflorus* 'Rubra'. Grows to 50cm tall, light green leaves with blood red tips.

FAMILY Liliaceae
GENUS *Liriope*
COMMON NAME Lilyturf or Turf lily

Appearance: Perennial, small, evergreen grass-like clumps; normally white or bluish flowers

Establishment: Prefers a sandy loam, grows well in sun or shade. Propagate by division.

Culture: Generally most hardy in mild temperate or cooler subtropical areas. Mulch in warmer climates to keep roots cool.

Cultivars

L. muscari—To 0.5m tall, purplish flowers.

L. spicata syn. *Ophiopogon spicatus*—Creeping plant, lilac to white flowers.

FAMILY Liliaceae
GENUS *Ophiopogon*
COMMON NAME Mondo grass

Appearance: Perennial herbs forming low grassy tussocks, spreading by underground roots which bud off to produce new tussocks. Occasionally produces small lily-like flower spikes.

Establishment: Plant any time. Grows in most soils, but will establish faster in a fertile, friable and moist soil. Grows in sun or shade. Excellent border or container plant. Propagate by division.

Culture: Not as invasive as bamboos, but it does spread, and can invade lawns, paved areas or virtually anywhere, although roots usually take a long time to start causing any serious damage. Responds well to watering and feeding, but will tolerate a fair degree of neglect.

Cultivars

Four species exist; cultivated varieties include:

O. intermedius—To 15–20cm tall. Green, glossy, recurved leaves, flowers similar to lily of the valley, white with a purplish tinge.

O. intermedius 'Argenteo Marginatus'—To 15cm tall; green foliage with pale cream margins.

O. japonicus (Common mondo grass)—To 20cm tall. Dark green, arching leaves, pale purple flowers. Other forms of this species include one with dark purplish foliage, and a dwarf (to around 7cm tall).

FAMILY Poaceae
GENUS *Pennisetum* syn. *Grymnothrix*
COMMON NAME Fountain grass

Appearance: Graceful, hardy, perennial grass with delicate plumes. Used in dried flower arrangements.

Establishment: Needs warm, preferably sunny position. Will grow in milder areas provided protected from extreme cold (does not do well in shade). Generally adapts to most soils, but prefers moist, drained and fertile soil. Propagates readily from seed. May also be divided.

Culture: *P. villosum* is often (but not always) treated as an annual, but *P. latifolium* is normally a permanent planting. Can become invasive in some hot climates.

Cultivars

P. latifolium—1.5 to 2m tall

P. villosum—To 0.4m tall.

FAMILY Poaceae
GENUS *Phyllostachys*
COMMON NAME Bamboo

Appearance: Spreading clump-forming bamboo. In hot areas it tends to grow more upright and taller; in temperate areas it is lower and more arched. Clean culms fall quickly. Rarely flowers.

Establishment: Grows well in temperate or subtropical climates; some also do well in the tropics. Generally very hardy and adaptable, but prefers full sun. Propagate by division.

Culture: Generally very easy, requiring little attention beyond controlling spread or watering in prolonged dry periods.

Cultivars

Widely grown species include:

P. aurea (Golden bamboo)—To 7m tall, thick yellowish stems.

P. nigra (Black bamboo)—To 9m tall; new stems are greenish, aging to black.

FAMILY Poaceae
GENUS *Saccharum*
COMMON NAME Sugar cane

Appearance: Large, spreading, bamboo-like grassy clumps to 4m or taller.

Establishment: Grows easily throughout the humid tropics. Likes moist, fertile soil, but generally adaptable. Propagate by cuttings.

Culture: Responds to feeding and watering while growing rapidly.

Cultivars

There are 12 species. *S. bengalense* is sometimes grown as a decorative plant; *S. officinarum* (Sugar cane) is the widely grown commercial plant, *S. sinense* (Chinese sweet cane) is sometimes cultivated for syrup.

5

Conifers

Many conifers can be grown successfully in the sub-tropics and tropics if you are careful about what varieties you grow, and how they are grown. Some conifers, in fact, are native to tropical and subtropical areas (eg. some *Araucaria, Agathis, Callitris* and *Pinus*).

The major problems with growing a wide variety of conifers in the sub-tropics or tropics are:

- **Roots getting too hot and dry**
  This can often be overcome by thick mulching, irrigation, and planting on cooler aspects, such as a slope facing away from the sun.
- **Foliage suffering from extreme humidity**
  This can sometimes be overcome by spacing plants further apart, pruning to prevent foliage becoming too thick, planting in open areas where there is plenty of air movement around the plants, and by avoiding overhead watering.

Generally, conifers should not be pruned hard, though their are exceptions. In humid climates a light, frequent pruning is important to keep a good shape, maintain ventilation and minimise the chances of any die-back.

FAMILY Araucariaceae
GENUS *Agathis*
COMMON NAME Kauri

Appearance: Tall, evergreen rainforest trees, with generally limited canopy spread—perhaps no more than 25% of the height— with entire leathery leaves.

Establishment: Prefers moist, highly organic soils. Feed and water young plants to ensure good growth. Protect young plants from frost or strong winds.

Culture: Mulch and water during extended dry periods. Valuable timber trees.Propagate by fresh seed sown at around 20–30°C.

Cultivars

A. australis (Kauri pine)—New Zealand native tree, to 50m tall.

A. palmerstonii (North Queensland kauri)—To 40m tall.

A. robusta (Queensland kauri)—To 40m tall.

FAMILY Araucariaceae
GENUS *Araucaria*
COMMON NAME Bunya Bunya, Monkey puzzle, Norfolk Island pine.

Appearance: Tall evergreen tree, stiff, small leaves arranged in whorls around stems, old trees often spreading with flattened top. Foliage is commonly dark green in colour.

Establishment: Does best in well drained fertile soil, in frost-free climates. Protect young plants but, once settled, many species will tolerate wind. Though many grow in the tropics, some will also adapt to colder conditions. Several species grow well in southern Australia. Can be faster growing than some other conifers if conditions are good. Propagate by fresh seed sown in well drained medium. Kept moist and at around 24–29°C.

Culture: Responds well to water and mulch in dry conditions. Can suffer from mealy bug and scale, otherwise few pests. Most species produce valuable timber. Large cones and prickly leaves shed from some species can be a nuisance. Can be prone to some diseases such as blight, canker and gall.

Cultivars

A. araucana (Monkey puzzle tree)—To 30m tall, from Chile.

A. bidwillii (Bunya Bunya)—To 35m tall, from subtropical mountain forests of Australia; seeds are edible.

A. columnaris (New Caledonia pine)—To over 60m tall.

A. cunninghamii (Hoop pine)—To 50m tall, from Australia.

A. heterophylla (Norfolk Island pine)—To 60m tall, often used in coastal plantings.

FAMILY Cupressaceae
GENUS *Callitris*
COMMON NAME Australian native cypress

Appearance: Large shrubs and small trees with graceful, fine foliage; commonly column-shaped with foliage extending close to ground level.

Establishment: Some, but not all, species suit hot climates. Generally faster growing than the average conifer, many tolerate extreme climates (eg. hot dry inland, frost, etc), but respond well to feeding and watering while establishing. Prefer well drained soils.

Culture: Once established, require little attention. Propagate easily from seed. Prone to several pests including aphids, caterpillars, scale, and diseases such as canker and needle blight.

Cultivars

C. collumellaris (Coastal cypress)—To 20m tall, blue-green foliage, pyramid shape.

C. endlicheri (Black cypress pine)—Grows 5–20m tall, green to blue-green foliage. Better for hot, dry inland areas than for wet tropics.

C. macleayana—Grows 10–30m tall. Does best in high-rainfall areas and well drained soils.

FAMILY Cupressaceae
GENUS *Cupressus*
COMMON NAME Cypress

Appearance: Evergreen trees or shrubs. Leaves are scale-like, hugging stems, typically conifer-like in scent and appearance.

Establishment: Most species not suited to the hot climates, but there are exceptions (see below).

Culture: Avoid overhead watering. Mulch annually to keep roots cool. Will propagate by seed or hardwood cuttings.

Cultivars

C. arizonica (Arizona cypress)—To 12m tall, pale to blue green foliage. Most suited to hot dry climates, but may tolerate some humidity. Several cultivars exist with variations in shape, size and foliage colour.

C. bakeri (Modoc cypress)—To 25m tall, narrow crown, grey-green foliage. May grow in subtropical areas.

C. glabra (Smooth-barked Arizona cypress)—To 20m tall, conical shape to 10m diameter. Tolerates heat or cold, but not wet tropics.

C. macrocarpa (Monterey cypress)— To 20m tall, fast-growing, spreading canopy perhaps 15m diameter, tolerates coastal winds. Successfully grown on some subtropical coasts.

C. sempervirens (Pencil pine)—Grows 15–25m tall, tolerates heat, but not extended periods of high humidity. Better for hot, dry inland areas than wet tropics, but is grown successfully in subtropical coastal Australia.

C. sempervirens 'Swanes Golden'—To 15m tall, narrow, column-shaped, golden foliage. Grows successfully on the Gold Coast in Queensland.

FAMILY Cupressaceae
GENUS Hybrid X *Cupressocyparis*
COMMON NAME Hybrid cypress

Note: These plants originated as a cross between *Cupressus macrocarpa* and *Chamaecyparis nootkatensis.*

Appearance: There are variations in size, colour and shape, but these are generally very large trees to 30 or 40m tall, with upright or columnar habit.

Establishment: Fast-growing and adaptable to a wide range of soils and climates. Proven throughout temperate climates and mountains and coastal parts of subtropical Australia. Tolerates wind and, once established, dry periods.

Culture: Responds well to fertiliser, irrigation and good drainage. Avoid overhead watering.

Mulch annually to keep roots cool. Propagate by cuttings.

Cultivars

C. 'Castlewellan Gold'—To 25m tall, bright yellow foliage.

C. 'Naylors Blue'—To 30m tall, blue green foliage, upright slender habit.

FAMILY Cupressaceae
GENUS *Juniperus*
COMMON NAME Juniper

Appearance: Some trees but mainly small plants, with needle- or scale-like foliage; often bushy or pyramid-shaped.

Establishment: Grow in most soils, but best in friable, well drained, fertile soil. Under good conditions growth can be relatively fast. Most species not suited to warmer climates, but some can grow well in subtropical areas, including south-east Queensland, and may even extend to high altitudes in the tropics. Propagate by cuttings.

Culture: Avoid overhead watering. Mulch annually to keep roots cool.

Cultivars

J. chinensis—Tree to 20m but with many smaller cultivars. Generally not suited to tropics but may be grown in cooler (eg. mountain) locations in the sub-tropics.

J. conferta—Prostrate to 3m tall, dense foliage. Very hardy species native to exposed coastal sites in Japan.

J. communis var. *compressa*—Columnar shaped, to 1m tall, dense foliage.

J. x *media*—A hybrid of *J. sabina* and *J. chinensis*, to 3m tall and 4m diameter, often smaller. Foliage varies between different cultivars.

J. sabina—Hardy, vigorous species to 2m tall and 5m diameter. Better for hot dry inland areas than for wet tropics.

FAMILY Pinaceae
GENUS *Pinus*
COMMON NAME Pine tree

Appearance: Generally large trees, needle-like foliage, and producing large 'pine cones'.

Establishment: Adapt to most ordinary soils, but prefer reasonable drainage. Most species not suited to the tropics, but there are exceptions (see below). Propagate by seed.

Culture: Feed annually, and water during extended dry periods. Common timber trees in many temperate and subtropical areas. Good windbreak plants. Susceptible to white ants; control measures may be needed in the tropics. Prone to a number of fungal diseases which may be exacerbated by a wet tropical climate.

Cultivars

P. canariensis (Canary Island pine)—Grows 20–30m tall. Better in drier rather than humid climates; will grow in dry inland tropics.

P. caribaea (Cuban pine)—To 30m tall, conical shape. Grows in humid lowland tropics, tolerates heat and humidity well.

P. merkusii—Sometimes grown in south-east Asia.

P. ornatum (Celebes pine)—Slender formed tree.

P. pinea (Stone pine, or Maritime pine)—To 25m, tolerates poor soils and salt winds. Can do well in hot, dry inland areas, but not very well in wet tropics.

FAMILY Podocarpaceae
GENUS *Podocarpus*
COMMON NAME Podocarp, Yellowwood

Appearance: Variable, from small shrubs to trees. Fibrous bark and simple, flat, shiny green leaves with a prominent mid vein. There are both tropical and temperate climate species.

Establishment: Species can grow well in mountains of tropics, wet tropics and sub-tropics. Prefers deep, well drained, fertile soil.

Culture: Smaller species can be trimmed as a hedge or pruned to shape. Propagate low bushy species by cuttings into moist sandy mix at around 24°C.

Propagate trees by slow germinating seed.

Cultivars

P. elatus—To 30m tall, light green leaves.

P. gracilior (African fern pine)—To 24m tall, green leaves with dark tip.

P. macrophyllus (Buddhist pine)—To 12m tall. Young shoots are light green, older leaves can be darker green.

P. polystachys—Small shrubby tree, dark green foliage.

P. rumphii—Slow-growing tree to 20m tall. Young growth is light green, older growth is dark green.

FAMILY Taxodiaceae
GENUS *Taxodium*
COMMON NAME Swamp cypress, Bald cypress

Appearance: Trees with deciduous, feathery foliage; green, changing to reddish tones in autumn if weather is cool enough.

Establishment: Hardy in subtropical climates, but avoid extreme winter cold. Performs best in deep, fertile and moist soils. Propagate by fresh seed, hardwood cuttings in shade, or layering branches.

Culture: Mulch and irrigate during extended dry periods.

Cultivars

There are only two species.

T. distichum—Grows 20–30m tall, native to south-west USA and Mexico.

T. mucronatum syn. *T. mexicanum* (Monterey cypress)—To 20m or taller, native to central and southern Mexico.

FAMILY Cupressaceae
GENUS *Thuja*
COMMON NAME Thuya

Appearance: Evergreen trees and shrubs, usually pyramid-shaped; small scale-like leaves.

Establishment: Most species not suited to the warmer climates, but there are exceptions. Cultivars of *T. orientalis* grow well in Brisbane and may do well in other tropical or subtropical areas. They generally prefer deep, moist, organic, well drained soils. Propagate by cuttings.

Culture: Avoid overhead watering. Mulch annually to keep roots cool. Prune every 3–6 months to maintain shape.

Cultivars

Five species, with many and varied cultivars. *T. orientalis* is perhaps the best for warmer climates.

T. orientalis 'Rosedalis'—Rounded shrub to 1m tall. Yellowish foliage in spring, becoming bluish through autumn.

T. orientalis 'Zebrina' syn. *T. orientalis aurea*—Grows 1–2m tall. Golden foliage in hot weather can turn bronze in cold conditions.

6

Fruit Plants

FAMILY Anacardiaceae
GENUS *Anacardium*
COMMON NAME Cashew nut

Appearance: Fast-growing evergreen tree to 12m.

Establishment: High temperatures needed. Low rainfall during flowering and harvesting essential. Tolerant of periods of drought. Drainage must be good. Propagate by seed and cuttings.

Culture: Adapted to a wide range of soil types and acidity. Apply fertiliser two or three times through the year, remove weeds from within the drip line. Mulch trees. Pests include bugs, stem borers, thrips, miners and scale. Anthracnose is the major disease.

Cultivars

A. occidentale (Cashew)—Leaves ovate, leathery with prominent veins, fruit is kidney-shaped nut on the base of the 'cashew apple' which is thin-skinned and edible.

FAMILY Bromeliaceae
GENUS *Ananas*
COMMON NAME Pineapple

Appearance: Evergreen rosette cluster of stiff leaves; flower and fruit develop on stalk rising from centre.

Establishment: Prefers well drained but moist organic soil. Often planted in raised beds on sloping ground. Needs full sun. Propagate by suckers or crowns.

Culture: Traditionally fed with shell grit and bone meal. Irrigate in dry weather.

Cultivars

A. comosus syn. *A. sativus*—To 0.7m tall, blue-green foliage.

A. comosus 'Variegatus'—Yellow variegation on leaves; cultivated as an ornamental, often in containers or baskets.

FAMILY Annonaceae
GENUS *Annona*
COMMON NAME Custard apple

Appearance: Evergreen trees and shrubs with fragrant foliage.

Establishment: Prefers a position with full sun; can tolerate light shade, but production may reduce. Propagate by cuttings, grafting and layering.

Culture: Irrigate if needed to keep soil moist during growing season.

Cultivars

Species grown include:

A. atemoya (Custard apple).

A. cherimola (Custard apple or Cherimoya) to 5m tall.

A. diversifolia (Ilama or Custard apple).

A. muricata (Soursop), to 3m tall.

A. reticulata (Custard apple or Mamon) to 5m tall.

A. squamosa (Custard apple or Sweetsop) to 6m.

FAMILY Moraceae
GENUS *Artocarpus*
COMMON NAME Jackfruit

Appearance: Evergreen trees with large green or reddish leaves.

Establishment: Prefers a shady, moist protected position. Grows and fruits best in tropical districts. Propagate by stem cuttings in winter, or suckers any time, or by seed.

Culture: Prune for shape, irrigate in dry weather. Fertilise annually.

Cultivars

A. altilis (Breadfruit)—To 16m tall, with large, deeply lobed leaves. Fruits may weight up to 4kg.

A. communis (Breadfruit)—Fast-growing, large tree with large round green seedless fruit.

A. heterophyllus (Jackfruit)—Slow-growing to 16m tall, with very large elongated, segmented, green and yellow fruit.

A. hypargyraea (Kwai muk)—Small evergreen tree with small round yellow fruit. Slow to fruit.

Other species cultivated include: *A. anisophyllus* (Entawak); *A. elasticus* (Tekalong); *A. kemando* (Paadau); *A. nitidus* (Selanking); *A. odoratissimus* (Marang); *A. sarawkensis* (Pingan) and *A. sericicarpus* (Pedalai).

FAMILY Oxidalaceae
GENUS *Averrhoa*
COMMON NAME Five corner fruit or Carambola

Appearance: Partially deciduous trees with pinnate leaves and semi-weeping appearance, especially when in fruit. Tends to keep foliage in tropics.

Establishment: Full sun to part shade tolerated. Protect from dry, strong winds, as branches may break. Propagate by cuttings, layering, graft or seed.

Culture: Water well in dry periods. Fertilise with complete plant blend or abundant animal manures. Prone to fruitfly attack, grasshoppers, and soil diseases if in wet soils.

Cultivars

A. bilimbi (Bilimbi)—Small tree with small pickle-like fruit of lower flavour quality.

A. carambola (Star fruit, Five corner fruit or Carambola)—Small tree to 5m with two fruiting seasons (summer and winter). Attractive and flavoursome fruit shaped as stars. Many varieties available.

FAMILY Solanaceae
GENUS *Capsicum*
COMMON NAME Chilli, Capsicum, Sweet pepper

Appearance: Small herbaceous plants normally grown as annuals. Green, yellow or red fruit used for culinary flavourings. Ornamental plant and fruit.

Establishment: Warm site in fertile, well drained soil high in organic matter. Full sun required. Prefer pH 5.5–6.6. High temperatures above 32°C with low humidity may cause flower and fruit drop. Temperatures below 16°C may cause poor fruit set. Propagate by seed.

Culture: Mulch soil well. Irrigate in dry periods. Avoid excess irrigation. Fertilise regularly with complete plant food or liquid solutions. Pests include aphids, fruitfly, nematodes and thrips. Diseases include anthracnose, bacterial wilt, leaf spot, mildew and virus attacks.

Cultivars

C. annuum (Capsicum, Sweet peppers)—Produces single fruits at leaf axils. Annuals. Fruits may be harvested 60–80 days after transplanting.

C. frutescens (Hot peppers)—Produce multiple fruits at leaf axils. Tend to be short-lived perennials. May tolerate less fertile soil, but requires more fertiliser and irrigation. Fruits are produced 80–100 days after transplanting.

FAMILY Caricaceae
GENUS *Carica*
COMMON NAME Paw Paw or Babaco

Appearance: Bare trunks topped with a cluster of large green leaves. Flowers emerge at top, followed by large edible fruits. Plants may bear male flowers only; or female only. Female flowering plants bear fruit, but some male plants are needed for cross pollination.

Establishment: Prefer fertile, moist soil, but generally adaptable to a wide range of soils. Under good conditions will mature and bear first crop within 1–2 years of planting. Can be propagated by cuttings or grafting.

Culture: Plants can rot in waterlogged conditions. Commonly propagated by home gardeners from seed. Once mature, male plants are thinned out, leaving only a few for cross pollination. Pests include fruitfly, grasshoppers, birds, viruses and a range of diseases.

Cultivars

C. papaya (Paw Paw or Papaya)—Grows 3–6m tall; available as male, female or bisexual fruiting plants.

C. pentagona (Babaco) Grows 3–6m tall. Deeper green leafed plant with parthenocarpic fruit (ie. only one plant needed). Best if grafted onto Paw Paw rootstock to avoid early root rot diseases.

FAMILY Rutaceae
GENUS *Casimiroa*
COMMON NAME Sapote

Appearance: Large to medium spreading trees with custard-like sweet fruit. Deciduous.
Establishment: Select a sunny warm site, best in tropics. Tolerant of sub-tropics, although fruiting is low. Propagate by grafting and seeds.
Culture: Well drained and fertile deep soil preferred. Fertilise with a complete plant food twice a year. Keep moist in dry periods.
Cultivars
C. edulis (White Sapote)—To 15m or taller, large yellowish fruit.
C. tetrameria (Yellow Sapote)—To 9m tall, large yellow fruit.
NOTE: Black Sapote is a different genus (*Diospyris*).

FAMILY Rutaceae
GENUS *Citrus*
COMMON NAMES Orange, Lemon, Lime, Pummelo.

Appearance: Mainly shrubby, evergreen trees. The mainly glossy, green leaves have a characteristic citrus scent when crushed. White flowers followed by yellow-orange fruits.
Establishment: Plant in well drained soil. Protect from extremes of wind or cold and keep moist while establishing. Most prefer slightly acidic soils. Provide full sun. Propagate by grafting, cuttings and seed.
Culture: Respond well to feeding and watering, but avoid waterlogging. Rots can attack the roots or base of the trunk. Keep thick mulch away from base, but otherwise use mulch in dry conditions to keep roots moist and cool. Prone to a range of pests and diseases.
Cultivars
C. aurantifolia (West Indian lime)—Small tree with small yellow sour fruit.
C. latifolia (Tahitian lime)—Small tree with medium green sour fruit.
C. limon (Lemon)—Medium tree with many varieties. Range of fruit flavours and fruiting periods.
C. madurensis (Calamondin)—Small tree with small round orange sour fruit.
C. paradisi (Grapefruit)—Medium tree with large yellow sweet-sour fruit, thick skinned.
C. maxima (Pummelo or Shaddock)—Medium tree with very large fruits.
C. reticulata (Orange, Mandarin)—Medium trees with variable fruit and sweet juicy flavours, depending on variety.
C. sinensis (Orange)—Sweet, juicy round fruits on medium to large trees.

FAMILY Rubiaceae
GENUS *Coffea*
COMMON NAME Coffee

Appearance: Evergreen shrubs or small trees, with reddish fruits that contain 2 seeds. The seeds are roasted to produce commercial coffee.
Establishment: Plant in full sun in the tropics and part shade in the sub-tropics.
Propagate by seeds or cuttings, preferably kept at around 27 to 30°C
Culture: Prefers a highly organic, fertile and moist soil. Fertilise annually with manure or fertiliser high in nitrogen. Susceptible to grasshoppers.
Cultivars
C. arabica—Grows 3–5m tall, with fragrant white flowers. Most commercial coffee comes from cultivars of this species.
C. liberica (Liberian coffee)—To 6m or taller; fragrant white flowers.

FAMILY Araceae
GENUS *Colocasia*
COMMON NAME Taro, talo, dalo,

Appearance: Edible tuber which produces a plant with large green elephant ears-like leaves. No trunk is produced, only blades of leaves wrapped around each other, which produce an upright habit. Other plants known as Taro, some also producing edible roots, include *Cyrtosperma, Alocasia* and *Xanthosoma*.
Establishment: Likes a frost-free area, prefer-

ably in the wet tropics or sub-tropics. Propagate by tuber cuttings or corms.

Culture: Prefers fertile, deep soil with good drainage. Abundant moisture required. Mulch heavily. A complete fertiliser application will aid tuber and leaf production (try 6:9:8 ratio). Few pests except taro leaf blight, nematodes, root rot. Tubers may be harvested 6–13 months after planting.

Cultivars

C. esculenta var. *escultenta* (Dasheen) — True taro, grows to 1m, produces a large central tuber.

C. esculenta var. *antiquorum* (Eddoe) — Dry land taro, produces many small tubers all clumped together.

FAMILY Ebenaceae
GENUS *Diospyros*
COMMON NAME Persimmon or Sapote

Appearance: Deciduous or evergreen shrubs or small trees, with very juicy fruit.

Establishment: Adapts to most ordinary soils. Some are hardy to warmer temperate climates if protected from extreme cold. Propagate by root suckers, budding or grafting.

Culture: Fertile soils preferred and regular applications required of a complete fertiliser mix. In some parts of the world, traditionally pruned as a vase system (like apple trees). Susceptible to fruitfly and fruit-piercing moths.

Cultivars

D. armata—To 6m tall, yellow fruits.

D. digyna (Black Persimmon, Chocolate pudding fruit)—Lush growing, small tree to 6m with glossy leaves and large green fruit. Pulp of fruit turns black when ripe.

D. discolor (Mabolo, Velvet apple)—Medium tree with large fruit covered with reddish-brown hairs. Eaten fresh.

D. kaki (Oriental persimmon)—Slow-growing, medium tree, with large yellow orange seedless sweet fruit. Deciduous.

D. virginiana (Persimmon)—Medium growing, deciduous tree with orange coloured sweet fruit.

NOTE: White Sapote is in the genus *Casimora* and is quite different to this group.

FAMILY Bombaceae
GENUS *Durio*
COMMON NAME Durian

Appearance: Evergreen tree. Fruit is large and spikey with a smell that offends many people, however the flavour is considered by some to be exquisite.

Establishment: Full sun to part shade in a protected site. Not tolerant of frost or drought. Propagate by seed or grafting.

Culture: Deep well drained acid soils. Diseases include canker, anthracnose and charcoal rot. Pests include fruit borers.

Cultivars

The commonly cultivated species is *D. zibethinus* (Durian). It is a fast-growing evergreen tree with large fruit covered with sharp protuberances, sweet and aromatic. May be partially deciduous in cooler areas. In various parts of the tropics, other species are cultivated including: *D. dulcis* (Red durian), *D. graveolens* (Red-fleshed durian); *D. kutejensis* (Lai); and *D. oxleyanus* (Beludu).

FAMILY Clusiaceae
GENUS *Garcinia*
COMMON NAME Mangosteen or Kandis

Appearance: Large tree to 15m, although very slow growing. Leaves usually thick.

Establishment: Strictly lowland, moist tropical climates needed to obtain significant fruit production. Propagate by 6–8cm semi-hardwood cuttings kept at around 24°C, or by seed.

Culture: Prune for shape only. Difficult to transplant trees of any size, otherwise little care required.

Cultivars

G. livingstonei (Imbe)—Slow-growing, small tree with separate male and female plants. Small round orange fruits with one seed.

G. mangostana (Mangosteen)—Regarded as the queen of fruits. Small (4–7cm) dark fruit with edible white sweet pulp. May take 8–15 years to bear fruit.

FAMILY Sapindaceae
GENUS *Litchi*
COMMON NAME Lychee

Appearance: Large, spreading tree usually only growing to 6m due to graft production. Glossy, light green leaves with terminal pendulous clusters of red fruit. Fruit is semi-translucent, succulent, with a large black seed.
Establishment: Select a sunny, well drained site protected from strong sea winds and frosts. Enjoys areas of warm summer humidity and cool, dry winters. Add compost to soil when planting. A cold period important to initiate flowering. Propagated by grafting to obtain quicker production, or by layering.
Culture: Water in dry periods. Mulch well to the drip line. Fertilise at least twice a year for established plants. Lychees are sensitive to excess nitrogen. Prone to erinose mite, grasshoppers, fruitfly, and a number of other diseases and pests.
Cultivars
L. chinensis (Lychee)—To 12m tall, attractive tree with edible fruit similar to grapes. Many varieties are available.

FAMILY Proteaceae
GENUS *Macadamia*
COMMON NAME Macadamia nut
Appearance: Large upright trees with spiny or smooth leaves. Pendulant flowers in racemes bear green husks that mature to expose brown, hard nut inside. Edible kernel is white.
Establishment: Provide a freely drained site in full sun to part shade. Plant grafted trees to obtain nut production sooner. Best if grown in frost-free areas. Provide protection from strong winds. Propagated by grafting or seed.
Culture: Responds to annual feeding. Unlike other proteaceous plants, has recently been found to like some feeding with phosphorus fertilisers. Prone to macadamia nut borer, bugs and rats. Diseases include Phytophthora rootrot.
Cultivars
M. integrifolia—To 20m tall, slightly wavy margins on leaves. The main source of commercial macadamia nuts.
M. tetraphylla—To 12m tall, sharply toothed leaves, sometimes used to produce commercial nut crops.

FAMILY Anacardiaceae
GENUS *Mangifera*
COMMON NAME Mango
Appearance: Large, dense tree with scented, deep-green leaves. Flower sprays produced in spring followed by fruit which mature to green-red. Large seed with fibrous fleshy pulp.
Establishment: Generally a tough plant, although added preparation will improve production. Tolerant of most soil types. Add compost and fruit tree fertiliser to site. Water well. Poor salt and frost tolerance. Propagated by seed or grafting.
Culture: Fertilise twice a year. Benefits from occasional sprays of seaweed solution or fish emulsion. Prune to control height and width. Only prune after fruit are harvested. Prone to fruitfly, anthracnose, root rots, scale, etc. Mangoes tend to be biennial bearing (ie. alternating between heavy crops one year, and little the next).
Cultivars
M. indica (Mango)—The main species grown. It is the common mango of commercial importance. Varieties now offer range of fruit sizes, colour and size of tree.
M. odorata (Kuwini or Wangi) is a large tree bearing large oval green fruit with sweet orange pulp. Several other species occasionally grown for fruit are: *M. caesia* (Belunu), *M. longipes* (Mango Ayer), *M. pajang* (Bambangan) and *M. quadrifolia* (Baab).

FAMILY Euphorbiaceae
GENUS *Manihot*
COMMON NAME Cassava
Appearance: Tall bamboo-like stems bearing lobed leaves and large tuberous root structures. Leaves are used as a cooking vegetable. Roots are cooked as a staple diet in many countries.
Establishment: Prefers a wind-protected site in full sun to part shade. Known to tolerate drought but sensitive to frost. Propagate by cuttings.
Culture: Adaptable to a wide range of soils from poor sand to heavy clay, but will not tolerate waterlogging. Fertilise to maximise leaf production and nutritional value. Prune if too

tall or to maintain shape. Pests include mites. Leaf spots and mosaic disease are recorded in some countries. Harvest tubers 6–24 months after planting. Leaves may be harvested continually.

Cultivars

M. dulcis (Sweet cassava)—Edible roots are generally free of poisonous acid found in *M. esculenta*.

M. esculenta syn. *M. ultissima*—Main commercial crop plant. Leaf shape is different to *M. dulcis*. Variegated leaf form popular as an ornamental garden plant. Edible roots contain bitter prussic acid which is destroyed in cooking.

FAMILY Sapotaceae
GENUS *Manilkara* syn. *Sapota* or *Achras*
COMMON NAME Sapodilla

Appearance: Evergreen tree, usually large, with milky sap, thick leaves and clusters of small whitish flowers. Grown both for latex production and as a fruit.

Establishment: Prefers well drained, friable, slightly alkaline soil. Will grow in wet or dry tropics, but prefers humidity. Young trees are frost-tender. Propagated by seed, budding, grafting. Layering used in India.

Culture: Responds to fertile conditions. Few major pests, though beetles and caterpillars are occasionally a problem. Seedling trees can take up to 8 years to bear; grafted trees can bear fruit in 2 years.

Cultivars

M. bidentata—To 30m tall, cultivated for latex or wood, rather than fruit.

M. zapota (Sapodilla)—Seedling trees to 30m tall; grafted varieties may only grow a fraction of this height. A hardy tree with glossy green leaves, the most popular species.

FAMILY Musaceae
GENUS *Musa* syn. *Ensete*
COMMON NAME Banana, Plantain

Appearance: Clump-forming, herbaceous plants with underground rhizomes which produce tall pseudostems with very large leaves. Starchy, edible fruit popular for humans and animals. Fruit are generally seedless.

Establishment: Needs sunny position, humid environment. Adapts to most soils (but avoid waterlogging). It is illegal to cultivate some types of bananas in some places (eg. Australia) without a permit. Such regulations are in force to control the spread of disease. Protect from wind damage. Propagate by suckers/offsets.

Culture: After fruiting, cut off old shoots and allow new growth from base to grow as a replacement. Responds to feeding at this time. Responds to watering in dry periods. Fruit are harvested when still green but ridges become more rounded. Disease such as Bunchy top and Panama disease require plants to be destroyed. Other problems include Sigatoka disease, banana root borer, nematodes.

Cultivars

The scientific naming of bananas is confusing. Some books refer to *Musa ensete* as the common banana, however others classify this as being *Ensete ventricosum* (belonging to a different genus).

M. acuminata—To 5m or taller, includes many named cultivars including Ladyfinger banana.

M. x *paradisiaca* (Plantain)—To 10m tall, very large fruits. Many named varieties are cultivated.

FAMILY Sapindaceae
GENUS *Nephelium*
COMMON NAME Rambutan

Appearance: Medium tree similar to lychee in appearance and habit. Glossy, green leaves and pendulous fruit either red or yellow.

Establishment: Prefers a tropical protected site with deep, fertile soil and sunny location. Propagated by seed, layering or grafting.

Culture: Regular irrigation important, although a short dry period of 3 weeks prior to flowering may improve flowering.

Cultivars

N. lappaceum (Rambutan)—Evergreen, medium tree to 25m tall, with sweet white pulp. Will not tolerate cold conditions.

N. mutabile (Pulasan)—Evergreen, large tree with small oval fruit with translucent flesh.

Other edible species cultivated include: *N. maingayi* (Lait), *N. melanomiscum* (Melajan) and *N. xerospermoides* (Parih).

FAMILY Lauraceae
GENUS *Persea*
COMMON NAME Avocado

Appearance: Large, evergreen shrub or tree with glossy green leaves. Fruit either green or black, smooth or rough. In some, male flowers open in the morning and females in the afternoon, and others vice-versa. These are classed as A and B pollinators.

Establishment: Best planted in spring or summer. Needs deep top soil, excellent drainage, but moist soil. Roots develop best at soil temperatures between 20 and 25°C. Later fruiting varieties (eg. Hass) do better in areas which may suffer cold winters. Tree guards are often valuable to protect plants during establishment. Propagate by grafting selected cultivars onto seedlings.

Culture: Control weeds around tree base. Root disturbance to the tree is detrimental. Fertilise frequently and lightly during periods of growth (liquid fertiliser or rotted manure). Remove terminal buds as tree develops to force branching. Frosts at flowering time will cause fruit drop. Susceptible to root rots, anthracnose and leaf and stem spots, also fruitfly attack.

Cultivars

P. americana (Avocado) — Evergreen tree to 12m tall. Many varieties now produce range of fruit shapes and are available as small or tall trees. Plant mixtures of A and B pollinators to maximise fruiting.

FAMILY Myrtaceae
GENUS *Psidium* (*Guava*)
COMMON NAME Guava

Appearance: Evergreen shrubs or small trees with yellow or red round or pear-shaped fruit. Leaves generally glossy green. Flowers typically small and white. Fruit bear many small seeds.

Establishment: Prefers good drainage, otherwise relatively hardy. Full sun preferred. Propagate by cuttings.

Culture: Fertilise and water regularly to promote prolific fruiting. Prune annually for shape only. Foliage is best kept dry when fruit is nearing maturity. Prone to fruitfly, sooty mould, scale, mealy bugs, etc.

Cultivars

P. araca (Araca)—Slow-growing evergreen small tree with small yellow firm fruit.

P. cattleianum (Strawberry guava)—To 7m tall, white flowers, reddish purple fruits.

P. guava (Guava) To 10m tall, yellow or red fruits. Fast-growing tree.

FAMILY Annonaceae
GENUS *Rollinia*
COMMON NAME Amazon custard apple, Rollinia, Biriba

Appearance: Tropical trees or shrubs with simple leaves.

Establishment: Plant when soil is wet, mulch lightly and keep soil moist for a settling in period. Well drained but moist soil is preferred. If necessary, provide protection from wind, frost, drought, salt and flood. Propagate selected cultivars by grafting onto seedlings.

Culture: For best results keep roots moist but avoid overwatering, which can cause root diseases. Avoid thick mulch around base of trunk (can cause collar rot). Responds to frequent light feeding. After harvest, prune any excessively long slender branches.

Cultivars

R. deliciosa (Amazon custard apple)—Grows 5–12m tall, fast-growing, large yellow fruit with brown protuberances and white pulp.

R. pulchrinervis—Tree with rusty-coloured new growths, and edible but poor-quality fruits.

FAMILY Sterculiaceae
GENUS *Theobroma*
COMMON NAME Cacao, Cocoa

Appearance: Slender, semi-deciduous tree with large hanging green pods that mature to tones of red. Brown to purple beans used to produce cocoa and chocolate. One species only, but three groups with distinct characteristics.

Establishment: Requires tropical climate, conditions. Protect from dry wind, salt spray, drought, frost and flooding. Tolerates rainfall of 1000mm and above. Optimum average daily temperature is 25.5°C with daily fluctuation not more than 10°C. Plant in a protected, shaded site. Propagate by seed or cuttings.
Culture: Deep, friable, fertile soil required. Remove weeds from around base of cacao plants. Mulch well. Irrigate in dry periods to ensure even distribution of water throughout the year. Fertilise to improve yield. Lightly prune during warmer months to maintain a shape conducive to harvesting. Aim for a well balanced canopy. Pods mature 150–180 days after fertilisation depending on variety. Prone to swollen shoot disease (virus), black pod disease, charcoal rot, and a range of other diseases. Major pests include mealybugs, borers, caterpillars and others.

Cultivars

T. cacao—The only species. Large elliptical leaves which emerge light green to pink. Small insignificant flowers.

7

Climbers

FAMILY Apocynaceae
GENUS *Allamanda*
COMMON NAME Golden trumpet

Appearance: Hardy, fast-growing, rambling shrubs or loose climbers, with long, pliable stems to 10m or more, and glossy, elliptical leaves to 15cm long in whorls. Flowers are large, colourful, five-petalled, trumpet-shaped, usually yellow, to 12cm across, carried in terminal clusters, sometimes followed by spiny, burr-like seed pods.

Establishment: Requires a sunny spot. Thrives in poor soils, but prefers moist, well drained soils. Minimum temperature of 13° celsius. Cuttings of old and new wood in spring or summer.

Culture: Keep dry during autumn season. Train to produce more lateral growth, which keeps flowers at a more visible height. Pruning is generally done in early spring. Prone to scale, red spider, mealy bug and thrips.

Cultivars:

A. cathartica—Large yellow trumpet flowers.

A. cathartica 'Hendersonii'—Orange-yellow blooms.

A. cathartica 'Grandiflora'—Pale-yellow blooms, not as abundant as other varieties.

A. cathartica 'Schotti'—Yellow blooms with brown-striped throats.

A. neriifolia—Large yellow trumpet-like flowers with orange-striped throats.

A. violaceae—Similar habit and appearance to *A.cathartica*, but hosting rose-pink flowers.

FAMILY Polygonaceae
GENUS *Antigonon*
COMMON NAME Coral vine

Appearance: Slender, vigorous, evergreen to semi-deciduous perennial climber with tendrils. Broad to ovate pale green leaves to 12cm long, with a heart-shaped base with slender tip. Flowers profusely over an extended period through spring and late summer, and has red, pink, white or yellowish blooms to about 2cm wide, in racemes that terminate in a tendril.

Establishment: Prefers partial shade and deep, moist, but well drained soil. Needs protection from hot, dry winds. Propagate by seed or cutting.

Culture: Prone to red spider, thrips and scale. Good bee-attracting plant.

Cultivars

A. leptosus—Climbing to about 12m, profuse, pink blooms, edible tubers.

A. leptosus 'Album'—White flowering form.

A. guatimalense—Similar to *A. leptosus*, but with thicker, hairier leaves.

FAMILY Aristolochiaceae
GENUS *Aristolochia*
COMMON NAME Calico flower or Birthwort

Appearance: Fast-growing, vigorous climber with large, fleshy underground roots or rhizomes, and triangular to heart-shaped leaves. Unusual purple, white and brown-mottled inflated flowers with a long 'tail', followed by seed pods like small hanging baskets.

Establishment: Prefers rich, moist soil. Will flower in sun to full shade. Propagate by cutting, layering or seed.

Culture: Responds well to feeding, and watering in dry periods.

Cultivars

A. elegans—Vigorous, slender climber to 7m tall, with dense kidney- or heart-shaped foliage to about 8cm across. Wide, shallow heart-shaped bowl-like maroon flowers with white markings. Doesn't have a bad smell like some other cultivars.

A. grandiflora syn. *gigantea, gigas*—Giant purple and white flowers, reaching up to 25cm in length and width. Also has a long 'tail'. Flower scent can be offensive.

A. grandiflora 'Hookeri'—Similar flower with smaller tail.

FAMILY Apocynaceae
GENUS *Beaumontia*
COMMON NAME Herald's trumpet, Easter lily vine

Appearance: Twining, showy plant to about 6m tall in cooler areas, and up to 15m tall in warm, moist areas. Young shoots are softly hairy and rust-coloured, followed by large shiny ovate to oblong leaves with hairy, red-brown lower surfaces. Flowers are fragrant, white with green veins, lily-like trumpets to 20cm long.

Establishment: Prefers a sheltered, semi-shaded position and rich, moist, loamy soil and a dry winter to promote flowering. Propagate by cuttings.

Culture: Prune immediately after flowering. Later pruning will reduce next season's crop of flowers.

Cultivars

B. grandiflora—See description above.

B. jerdonianan—Similar to *B. grandiflora*, but smaller in most botanical features, differing in that it has red flower buds, a more abruptly expanded flower tube, and almost smooth leaf undersurfaces. Believed by many to be a variety of *B. grandiflora.*

FAMILY Bignoniaceae
GENUS *Bignonia*
COMMON NAME Cross vine

Appearance: Vigorous evergreen climber to about 16m tall, with compound leaves having 2 ovate to lance-shaped leaflets and a terminal tendril. Flowers are in clusters, 4–5cm long, foxglove-like, yellow-red, going from darker colour on the tip to lighter on the inside.

Establishment: Tolerates most soils, but prefers a deep, rich, moist, well drained soil. Can be grown in cooler climates, but with poor flowering. Propagate by seed (when available) and cutting.

Culture: Prune to control size and water in dry periods. Few pest and disease problems.

Cultivars

B. capreolata—See description above.

B. capreolata 'Atrosanguinea'—Longer, narrower leaves and darker flower colour.

FAMILY Nyctaginaceae
GENUS *Bougainvillea*
COMMON NAME Paper flower

Appearance: Mainly vigorous, woody vines. Some of the older varieties may reach up to 15m tall. What is commonly considered the flower of the bougainvillea are actually showy coloured bracts covering the true, tubular flower. Flowering season is from spring to late summer. Elliptical to ovate leaves are softly hairy beneath. Some cultivars may have spines. A wide variety of colours available. Extensive hybridisation has been carried out. The main parents are *B* x *buttiana, B. glabra* and *B. spectabilis.*

Establishment: Tolerant of most soils, but prefer rich soils. Dislikes waterlogged soils. Does best in full sun. Propagate by cuttings.

Culture: Can be grown in pots, left to climb or trained to a standard. Flowers best in dry conditions and is tolerant of reflected heat. Generally few pest and disease problems. Prune well after flowering to control growth.

Cultivars

B x *buttiana*—Wide variety of cultivars with mainly orange, pinks and red bracts.

B. glabra—Bracts are magenta to purple. There are numerous cultivars including:
'Alba'—pure white flowers.
'Formosa'—a strong grower with lilac flowers.
Magnifica which has deep purple bracts.
Magnifica trailii which has magenta bracts.
'Sanderana'—bright purple bracts and downy leaves.

B. spectabilis—Bracts are typically purple, pink or light red. Cultivars provide an extensive range of colours. Dwarf and variegated forms are also available.

FAMILY Bignoniaceae
GENUS *Campsis*
COMMON NAME Chinese trumpet creeper

Appearance: Vigorous, woody-stemmed, deciduous climbers to 15m; the stems may have aerial roots. Opposite, pinnate lance-shaped leaves, softly glabrous beneath. The showy clusters of tubular flowers appear in summer, and come in hues of reds and orange.
Establishment: Fertile soil and full sun preferred, but generally hardy. Propagate by root cutting, vegetative cutting, suckers, layering or seed.
Culture: Little attention required.
Cultivars
C. grandiflora syn *B. chinensis*—Flower opens orange, fading to apricot with red veins. Not a very secure root climber, and may require some support in windier positions.
C. grandiflora 'Thunbergia'—Orange flowers.
C. radicans—Bright scarlet base with orange trumpet flower opening in summer; very showy. A strong root climber and seldom needs support.
C. x *tagliabuana* (*C.grandiflora* x *radicans*)—Flowers are five-lobed red trumpets.
C. x *tagliabuana* 'Mme Galen'—A salmon flowering variety, it is the most common seen in cultivation.

FAMILY Asclepiadaceae
GENUS *Ceropegia*
COMMON NAME Fountain flower

Appearance: Slender twining plants. Most grow to about 1 metre long.
Establishment: Requires well drained soil and full sun. Suited to basket or pot.
Culture: . Propagate by seed or cutting. Keep soil nearly dry from late autumn to mid-spring. Minimum temperature of 10°C is required.
Cultivars
C. lineris—Flowers are tubular, in clusters of 3 to 5, green with purple stripes.
C. haygarthii—Unusual flower with pink-cream tube with purple spots, the base makes a 90 degree turn and flares open at the mouth, exposing a red knotted stigma.
C. sandersonii—The flower is mottled light and dark greens with a wide, five-petalled lobe opening. A thin membrane stretches across the lobes, creating a fountain effect.
C. woodii (String of hearts)—Thread-like pendulous stems with small heart-shaped, succulent leaves, with marbled upper surface and purple lower surface. Requires a protected, semi-shaded position, rich soil, and an occasional soaking, but otherwise infrequent watering.

FAMILY Vitaceae
GENUS *Cissus*
COMMON NAME Kangaroo vine, Grape ivy

Appearance: Fast-growing, small to medium-sized herbaceous or woody-stemmed, sometimes fleshy vines. Glossy deep green, simple or palmately compound leaves, most having tendrils. Small, greenish flowers followed by round black berries, not edible.
Establishment: Adaptable; will tolerate low light and low humidity. Good container plant. Commonly grown as indoor plant in cooler areas. Propagate by seed or cutting.
Culture: Prune freely as required, provide something to climb on. Red spider may be a problem.
Cultivars
C. antarctica (Kangaroo vine)—Good screening plant or for large hanging baskets. Dense, rich green, toothed foliage up to 15cm long.
C. antarctica 'Minima'—Slow-growing dwarf variety, excellent for hanging baskets.
C. discolor—Soft-wooded climber requiring rich, moist, well drained soil in a sheltered position with filtered light. Minimum temperatures of at least 15° C. Leaves velvety-green with silvery-white bands between the veins, the underside an attractive maroon-red.
C. hypoglauca—Flowers are small and yellow, followed by attractive round blue-black berries.

FAMILY Ranunculaceae
GENUS *Clematis*
COMMON NAME Virgin's bower

Appearance: Attractive, medium-sized climbers to 3m or more, with often very showy flowers. Some are evergreen, some deciduous.

They have compound leaves and use the midribs to support themselves. Flowers range in appearance from delicate star shapes to larger lobed petals. Pale white to purple colours.
Establishment: Prefers rich, moist, well drained soil. Most commonly grown species are temperate plants, but there are others that will grow in warmer climates. Propagate by layering in spring, cutting in summer, or by division.
Culture: Likes cool roots, but will tolerate sun on top growth; prefers a partially shaded location. Nematodes and rabbits can be problems.
Cultivars
- C. *afoliata*—Flowers are pale greenish yellow and fragrant, growing in clusters.
- C. *hookerana*—Evergreen climber from New Zealand. Very fragrant five-petalled star flower. Male flowers are white, about 5cm across, female flowers are smaller.
- C. *meyemiana*—Flowers are clusters of four sepalled petals in panicles.

FAMILY Verbenaceae
GENUS *Clerodendrum*
COMMON NAME Bleeding heart vine

Appearance: Fast-growing, twining, evergreen shrubs or vines reaching to 4 metres. Flowers are white with a red calyx, appearing heart-shaped and in clusters. Suckers readily.
Establishment: Prefers a well drained, fertile soil in a protected position with filtered light to semi-shade. Propagate by seed or cutting.
Culture: Mature plants can suddenly die unexpectedly. Mealy bug and scale may be problems. Can be transplanted fairly readily.
Cultivars
- C. *splendens*—see description above. Has attractive red flowers in spring and autumn.
- C. *thomsonae* syn. C. *balfouri*—a vigorous, evergreen climber with ovate to ovate-oblong leaves to 13cm long, and clusters of showy white flowers with a crimson-red central tube. It prefers rich, moist, but well drained soil, and a protected position.
- C. *thomsonae* 'Delectum'—has large clusters of rose-magenta flowers.

FAMILY Fabaceae
GENUS *Clianthus*
COMMON NAME Parrot's beak, Glory pea

Appearance: Thin, scrambling, shrubby plant to 3m. Pinnate leaves with 15 oblong leaflets.
Establishment: Needs well drained soil, otherwise fairly adaptable. Propagate by seed or cutting
Culture: Protect from cold wind or frost. Prune after flowering to keep bushy.
Cultivars
- C. *puniceus*—The flower is a very showy pea flower, similar to the Sturt Desert Pea (C. *formosus*), which has a beak tip, flared pea-centre and swept back petal. The flower is large, up to 8 cm, and brilliant red.
- C. *puniceus* 'Albus'—Cream-white flower.
- C. *puniceus* 'Roseus'—Rose shades.

FAMILY Fabaceae
GENUS *Clitoria*
COMMON NAME Butterfly pea

Appearance: Shrubby climbers; leaves are pinnate, leaflets mainly ovate to oblong. Pea-like flower with winged petals, similar in appearance to a small butterfly.
Establishment: Adaptable, will do well in moist semi-shaded positions. Easily grown from seed, or cuttings at around 27°C.
Culture: Water freely when putting on growth, reduce watering if growth slows. Few pest and disease problems.
Cultivars
- C. *heterophylla*—To 70cm tall, blue flowers.
- C. *ternatea*—To 4m or taller, attractive blue and white pea-shaped flowers. Double and white-flowered forms also grown.

FAMILY Bignoniaceae
GENUS *Clytostoma*
COMMON NAME Argentine trumpet vine, Trumpet flower

Appearance: Evergreen, shrubby climbers commonly to 5m long vines. Leaves are pairs of long ovate to elliptical leaflets with a terminal tendril. The flowers are five-lobed trumpets, lilac with purple streaks, although colour does vary.

Establishment: Prefers rich, moist, neutral to slightly acidic soil, and cool, dry winters. Needs support. Propagate by cuttings.
Culture: Water freely only when in growth. Keep compact by late summer pruning, removing up to 40% of the growth.
Cultivars
C. binatum—To 3m tall; flowers to 60mm long and wide, mauve with white throat.
C. callistegioides—To 5m tall; flowers to 75mm long and wide, lavender, streaked with violet.

FAMILY Combretaceae
GENUS *Combretum*
COMMON NAME Burning bush
Appearance: Showy plant with elliptical, opposite leaves on soft, hairy stems with spines. Masses of showy red spikes of 4- to 5-petalled blooms appear in midsummer, followed by clusters of interesting green seed pods. Shrubby, fast-growing, woody vines, often very large in size.
Establishment: Adaptable, preferring rich, moist soil in a sunny position. Easily propagated by seed.
Culture: Prune when young to keep under control. Scale may be a problem.
Cultivars
C. grandiflora—To 7m tall, with small, showy red flowers to 3cm long.
C. paniculatum—Very large vine with spiny stems and coral-red flowers, followed by small pink or orange fruit.

FAMILY Verbenaceae
GENUS *Congea*
COMMON NAME Shower of orchids
Appearance: Vigorous, shrubby climber with simple, entire, oval leaves. The flowers are similar in appearance to orchids, with white to lilac bracts and tiny white centres.
Establishment: Prefers a fertile, moist soil in a sunny position. Propagate by seed or hardwood cuttings.
Culture: It requires strong support (ie. tree or trellis). Water in dry periods; mulch and fertilise annually.
Cultivars
C. tomentosa—Large, spreading climber with a spectacular display of lilac to white flowers in spring, that can be cut for indoors.

FAMILY Moraceae
GENUS *Ficus*
COMMON NAME Fig
Appearance: Woody root-clinging vines with milky sap. Stiff mature stems can be shaped. Small globular, oblong or pear-shaped seed pods (figs), but these are not particularly showy. Grown mostly for foliage effects.
Establishment: Likes well drained soil, sun or part shade. Will take full shade in warmest areas. Propagate by cuttings.
Culture: Can become invasive, blocking pipes or causing structural damage to walls. Prune to control spread, and inspect for structural damage at least annually. Remove at the first sign of penetrating cracks in walls or other structures. Likes watering in dry periods.
Cultivars
F. pumila—Vigorous creeping/climbing plant. Young leaves small and attractive; as it matures large, less attractive leaves and heavy figs are produced. Regular pruning needed to get rid of these. If allowed to grow unchecked, the branches can become very heavy and can cause supporting framework to collapse. Can be grown in hanging baskets, and can be topiaried.
F. pumila 'Quercifolia'—Lobed leaves.
F. pumila 'Variegata'—Green and white variegation.

FAMILY Asclepiadaceae
GENUS *Hoya*
COMMON NAME Wax plant
Appearance: Small, twining, evergreen root climbers with simple, entire sometimes succulent or leathery leaves. Umbel-shaped clusters of fragrant flowers. Individual flowers are trumpet-shaped with five lobes.
Establishment: Prefers light shade and moist, well drained soil. Some make excellent container plants. Propagates easily but sometimes slowly by cuttings or layers.
Culture: Best flowering is often when the plant is potbound and not heavily fertilised. The

remains of old flowers should not be removed as new flowers arise from these.

Cultivars

H. australis—a succulent twiner with oval to orb-shaped leaves and white flowers with red at the base.

H. carnosa—umbels of waxy pink buds opening to pink stars with a red centre. This plant is very fragrant, especially at night.

H. carnosa 'Alba'—white flowering form.

H. carnosa 'Variegata'—white margined leaves.

FAMILY Convolvulaceae
GENUS *Ipomoea*
COMMON NAME Morning glory

Appearance: Generally vigorous twining creepers, with mainly ovate, occasionally lobed leaves, and funnel- or trumpet-shaped flowers in a wide variety of colours, particularly in different shades of reds, blues, purples and pinks. Some *Ipomoea* species, while being very attractive, naturalise readily and are potential weeds (eg. *I. alba, I. acuminata, I. tricolor, I. purpurea*).

Establishment: Adaptable; most prefer a warm, sunny spot and fertile, humus-rich soil. Propagate by seed or cuttings.

Culture: Prune and, if necessary, cut roots to control growth. Water in drought. Responds to mulch and fertiliser.

Cultivars

I. alba (*I. bona-nox*)—Dense foliage, milky sap, large, fragrant white flowers that open at night. Likes regular moisture and fertile soil. Widely naturalised in many tropical areas.

I. coccinea—An annual twining climber to 3m which requires support. It has ovate-cordate leaves to 15cm long. In late summer and autumn it has yellow-throated, striking scarlet flowers to 4cm long.

I. horsfalliae—Attractive, large, woody climber with clusters of rose-red to pale purple flowers to 7cm long and with white stamens. Best in hot, humid conditions.

I. horsfalliae 'Brigsii'—Masses of magenta-crimson flowers in spring and sometimes autumn. Prefers a sheltered position with morning sun.

FAMILY Oleaceae
GENUS *Jasminum*
COMMON NAME Jasmine

Appearance: Small to medium climber with shiny pinnate oval to lance-shaped leaves. The flowers, which are mostly fragrant, are in clusters at the tip of stems, with a fine tube opening to six petals.

Establishment: Very adaptable plants, growing in most soils. Likes a sunny to partly shaded position. Propagate by cuttings.

Culture: Most are hardy and relatively pest- and disease-free.

Cultivars

J. dichtomum—Scrambling, woody climber growing up to 12 metres.The flower tubes are white with red-tinged, white-lobed petals.

J. lineare—Pinnate, narrow leaves of 3 to 5 leaflets, very attractive foliage. Flowers are a dark pink bud opening to a white to pink bloom with eight thin petals. Nice fragrance.

J. grandiflorum—Evergreen, shrubby, scrambling climber. The flowers are white.

J. polyanthum—Popular, vigorous, scrambling climber that is deciduous in cooler areas, and self layers. It has very fragrant white, star-shaped flowers in spring.

FAMILY Apocynaceae
GENUS *Mandevilla*

Appearance: Woody twiners with opposite pairs of mainly simple leaves, milky sap, and trumpet-shaped flowers.

Establishment: Best in moist, rich, well drained soils in a warm, sheltered position. Propagate by seed in spring, or by cuttings.

Culture: Water moderately when growth rate slows. Prune shoots back to 2 or 3 buds from the base after flowering.

Cultivars:

M. x *amabilis* (*Dipladenia* x *amabilis*)—Woody, slow-growing twiner with very large, showy, funnel-like flowers that open pale pink and change to deep rose, over a long period, and open at night. Needs a protected position.

M. laxa (Chilean Jasmine)—A hardy, woody twiner, with heart-shaped leaves to 15cm long, and clusters of 5 to 15 fragrant white to cream-coloured flowers to 5cm long, in midsummer to early autumn.

FAMILY Rubiaceae
GENUS *Manettia*
COMMON NAME Brazilian firecracker vine

Appearance: Only one species widely cultivated. *Manettia inflata* syn. *M. bicolor* is a delicate, soft-stemmed, evergreen, perennial twiner to about 1.5m tall with pairs of rich, green, oval to lance-shaped leaves to 15cm long. Masses of small, red, tubular flowers with yellow tips.

Establishment: Prefers a shaded, protected position in a rich, moist, well drained soil. Otherwise hardy, growing out of doors from the tropics to temperate climates (eg. will grow in Melbourne). Propagates easily by cuttings.

Culture: Mulch and water in prolonged drought. Prune for shape. Prone to red spider infestation.

FAMILY Fabaceae
GENUS *Mucuna*
COMMON NAME Red jade

Appearance: Dense, evergreen, vigorous climber with magnificent flowers.

Establishment: Prefers a deep, fertile, moist soil in a warm, humid position. Propagate by seed.

Culture: If necessary, prune to shape. Frost-tender and needs support.

Cultivars

M. bennettii—To 12m tall, with long pendant clusters of large orange, red to scarlet pea-shaped flowers.

FAMILY Bignoniaceae
GENUS *Pandorea*
COMMON NAME Pandorea or Wonga Wonga vine

Appearance: Slow to medium-fast growing, attractive, evergreen climbers, with glossy, green, pinnate foliage, and terminal clusters of attractive tubular flowers.

Establishment: Hardy and adaptable, preferring, fertile, moist, well drained soil in full sun to filtered light. Propagate by seed or cuttings.

Culture: Excellent for trellis or covering fences. Responds to mulching. Prune to control shape or size only.

Cultivars

P. jasminoides—To 2m tall, with beautiful pink or white flowers. Several cultivars available, including a very attractive, more floriferous form, 'Southern Belle'.

P. pandorana—A more vigorous climber to about 6m tall, and fragrant cream, tubular flowers with a crimson or purple blush in the throat. Several cultivars with different coloured flowers available.

FAMILY Passifloraceae
GENUS *Passiflora*
COMMON NAME Passionfruit

Appearance: Fast-growing tendril climbers with attractive flowers, mainly in summer and autumn. Some produce tasty fruit.

Establishment: Adaptable and generally hardy. Prefer fertile, moist, well drained soils. Propagate by seed, cuttings, grafting, suckers (some).

Culture: Responds to feeding, watering and mulching. Do not disturb roots (avoid cultivation).

Cultivars

P. caerulea—Fast-growing, hardy climber with attractive blue flowers, suckers strongly.

P. edulis—The common edible passionfruit, it is a fast-growing, attractive climber. It may need to be replaced fairly regularly as the fruit can suffer from 'woodiness disease'.

P. laurifolia (Yellow granadilla)—Laurel-like leaves, and attractive blue flowers. Lemon-yellow to orange, soft-skinned, edible fruit, bears well in tropics but may not produce as well in subtropical areas.

P. ligularis (Sweet granadilla)—Yellow to purplish ovoid, edible, sweet-tasting fruit with white pulp.

P. quadrangularis—Large flowers followed by large, edible, thick-rinded fruit to 30cm long. Needs high humidity, and won't tolerate frost. Best in tropics, but will grow in a sheltered position in the sub-tropics. May need to be hand pollinated if humidity is low.

FAMILY Araceae
GENUS *Philodendron*

Appearance: Commonly a spreading shrub or climber, usually with large heart-shaped leaves (sometimes but not normally with indented or divided leaf margin). Leaf colour often green but can vary according to variety.

Establishment: Prefers filtered light to medium shade, but many will grow in heavier shade. Likes a fertile, well drained but moist organic soil, but adapts to other soils in moist humid climates. Most are hardy provided they are well drained, moist and protected from low temperatures (ie. 5° C). Some tolerate temperate climates if protected from frost. Most prefer humid tropical or subtropical climates. Propagate by cuttings any time.

Culture: Climbers may need tying up. Prune freely to maintain shape. Reduce watering if weather becomes cool. Overwatering and cool conditions can result in root rots.

Cultivars

Around 200 species including dozens of named varieties in common cultivation throughout both the tropics, and as indoor plants in cooler climates.

FAMILY Bignoniaceae
GENUS *Pyrostegia*
COMMON NAME Golden shower.

Appearance: One species commonly grown. *P. venusta* is a vigorous, fast-growing, evergreen, tendril climber to at least 10m tall. In spring the vine is nearly completely covered in masses of pendant clusters of 5–6cm long golden-orange flowers.

Establishment: Hardy, adaptable, preferring a sunny spot in fertile soil. Propagate by cuttings.

Culture: Prune after flowering to control size and shape. Water and mulch in extended dry periods.

FAMILY Araceae
GENUS *Scindapsis*
COMMON NAME Devil's ivy, Golden pathos

Appearance: Hardy climbers with commonly coloured or variegated heart-shaped leaves.

Establishment: As for *Philodendron*.

Culture: Climbers may need tying up. Prune freely to maintain shape. Reduce watering if weather becomes cool. Overwatering and cool conditions can result in root rots.

Cultivars

Commonly grown species include:

S. aureus syn. *Epipremnum aureum*—Fleshy clinging vine with dark green and yellow variegated leaves. Commonly grown species with several variations in cultivars available.

S. pictus—Clinging vine with thick leathery greenish leaves with silver blotching.

FAMILY Solanaceae
GENUS *Solanum*

Appearance: Vigorous, fast-growing, often poisonous climbers, with alternate, simple or dissected leaves, and single or clusters of star-shaped flowers similar to a potato's.

Establishment: Most prefer a sunny, protected position in a rich, moist, well drained soil. Propagate by seed and cuttings.

Culture: Prune for size or shape. Responds well to mulching or watering in dry periods.

Cultivars

S. jasminoides—Not an invasive climber to 5m tall. Small white or blue flowers in great profusion, over an extended period.

S. seaforthianum—To 5m tall, blue or purple flowers.

FAMILY Asclepiadaceae
GENUS *Stephanotis*

Appearance: One species commonly grown as a climber (*S. floribunda*). It is a vigorous twiner to 6m or more, with attractive, fragrant white flowers, that last well when cut.

Establishment: Prefers a rich, moist soil in a warm, sheltered position. Will do well in shade. Propagate by seed, cuttings and layers.

Culture: Prune to control size only.

FAMILY Araceae
GENUS *Syngonium*
COMMON NAME Goosefoot climbers, Arrowhead vine

Appearance: Elongated triangular heart-shaped leaves.

Establishment: As for *Philodendron.*
Culture: As for *Philodendron*. Remove tip growth if desired to keep plants bushy.
Cultivars
Approximately 20 species and many named varieties are commonly grown as container plants both in temperate (indoors) and warm climates. Also grow easily in the ground in hot climates.

FAMILY Bignoniaceae
GENUS *Tecomanthe*
Appearance: Slow-growing, evergreen twiner with pinnate leaves and clusters of attractive flowers that appear on the old wood.
Establishment: Prefers an organic-rich, moist soil, in a warm position with a moist atmosphere. Propagate by seed (spring) or cuttings (summer).
Culture: Responds to regular mulching and watering during dry periods.
Cultivars
T. dendrophylla—Pinnate leaves to 8cm long. Large, hanging, trumpet-shaped mauve-pink flowers with a paler throat, followed by pointed bean-like seed pods to 30cm long.
T. hilli and *T. speciosa* are others to consider.

FAMILY Acanthaceae
GENUS *Thunbergia*
COMMON NAME Black-eyed Susan
Appearance: Quick-growing annual and perennial climbers, with attractive blue, yellow, orange or white flowers.
Establishment: Adapts to varying conditions, but prefers a warm, sunny position. Propagate by seed, cuttings and layers.
Culture: Many species can become invasive, or roots can cause structural damage (some are banned in Queensland). Prune hard if need be to control top growth. Grow in pots to control root system.
Cultivars
T. alata—A small, very attractive climber to 3m tall, that is often treated as an annual. Showy white or orange-yellow 4cm long flowers with a dark purple throat. Prone to red spider attack.
T. grandiflora—Vigorous, hardy, fast-growing, woody-stemmed climber to 10m or more. Lavender to blue flowers to about 7.5cm wide with a yellow throat that has bluish veins. A white form is also grown.
T. gregorii—Similar to *T. alata* but flowers are orange, to 5cm wide, without the dark centre.
T. mysorensis (*Hexacentris mysorensis*)—A vigorous climber with spectacular, very large pendant racemes to 1.2m long, of yellow and red pea-shaped flowers in spring. Needs rich soil, humidity, and support.

FAMILY Apocynaceae
GENUS *Trachelospermum*
COMMON NAME Star jasmine
Appearance: Twining or clambering shrubby vine with milky sap, some with aerial roots, and masses of small, fragrant flowers.
Establishment: Adaptable, preferring fertile soil and regular watering. Protect young plants from wind or cold. Propagate by cuttings.
Culture: Requires little attention once established, apart from protection from severe weather, or pruning to control growth.
Cultivars
T. jasminoides—Hardy, large, evergreen, dense-foliaged twiner to 10m or more, that makes a good cover over embankments, on trellis and fences. Covered in spring with small, white, fragrant flowers to 2.5cm across, similar to Jasmine.

8

Low-growing Plants

This chapter deals with low-growing plants, including ground covers and small shrubby species generally less than around 1 metre tall, and grown for decorative purposes.

You may find plants here for any of the following applications:

- To fill in gaps between larger plants (eg. to cover the ground below palms).
- To provide a border to a path or garden bed, in front of larger plants.
- As pot plants for use indoors or out.
- As rockery plants, to hang over walls, or cover embankments.

You may find other plants to achieve similar purposes in the chapter on bulbs.

FAMILY Gesneriaceae
GENUS *Aeschynanthus*
COMMON NAME Basket plant, Lipstick plant

Appearance: Tender, trailing plants with generally attractive reddish, orange or greenish tubular flowers.
Establishment: Needs well drained but always moist, organic soil, Needs humid, warm climate. Prefers good light, but not direct sun. Ideal in baskets, pots, dead logs, etc. Best grown in pots outside of the tropics, with protection during winter.
Culture: Fertilise with slow-release pellets or liquid fertilisers. Trim plant to maintain a neat appearance. Propagate by cuttings.
Cultivars
A. lobbianus syn. *A. radicans*—Bright red flowers with yellow base.

FAMILY Araceae
GENUS *Aglaonema*
COMMON NAME Chinese evergreen

Appearance: Clump-forming perennials normally with variegated leaves, and occasionally an arum (spathe)-like flower.
Establishment: Prefers a shady position, highly organic but well drained soil, or good quality potting mix. Avoid extreme cold or wind. Does well in pots or in the ground in a moist position.
Culture: Water to keep soil/potting mix moist in dry periods. Prone to mealy bugs, spider mite, scale. Prune if plant becomes long and open. Cut low to encourage vigorous, bushy new growth. Propagate by stem cuttings.
Cultivars
There are 50 species and many named varieties, commonly grown as indoor plants in both temperate and tropical climates, or as a ground cover or border to shrubberies in hot, humid climates.

FAMILY Lamiaceae
GENUS *Ajuga*
COMMON NAME Bugle

Appearance: Rosettes of tender leaves emerge from trailing stems.
Establishment: Grows well in most soils. Very hardy, grows from tropics to temperate climates.
Culture: Mulch and water during periods of prolonged dryness. Hard to kill, but can wilt and die back if exposed to extreme wind, sun or dryness. Propagates very easily by division.
Cultivars
A. chia—To 20cm, yellow flowers. Tolerates hot, dry climate.

A. reptans—To 15cm tall. Blue to pinkish or white flowers. A dark-purple leafed variety is also commonly grown.

FAMILY Amaranthaceae
GENUS *Alternanthera*
COMMON NAME Joy weed, Copperleaf, Joseph's coat

Appearance: Shrubby, spreading perennials with colourful foliage.
Establishment: Adapts to most soil types, but responds best in fertile, friable soils. Best in full or lightly filtered sun. Plants last for years in the tropics; sometimes treated as annual bedding plants in temperate climates.
Culture: Prune to control size and shape. Water and mulch in dry weather. Propagates readily by cuttings or division.
Cultivars
A. dentata—To 50cm tall; green or coloured foliage.
A. dentata 'Rubiginosa'—To 50cm tall; purplish foliage.
A. ficoidea—Many named varieties exist varying in height and foliage colour. Leaves may contain shades of yellows, reds or green. Plant height varies from low to up to 40cm tall.

FAMILY Araceae
GENUS *Anthurium*
COMMON NAME Flamingo plant

Appearance: Clump-forming, heart-shaped, glossy leaves and large, attractive colourful spathe flowers.
Establishment: Requires moist, well drained soil, with high organic content; and humid atmosphere. Bright light required to encourage flower display.
Culture: Mist plants regularly to keep humidity up. Remove old leaves and spent flower stalks.Fertilise with slow-release pellets or liquid products. Susceptible to mealy bugs, mites and root rots. Propagate by division, or seeds kept at around 27°C.
Cultivars
A. andreanum—Ovate leaves to 25cm long sit at the ends of long stalks normally emerging from the base, though stems can eventually be more than 30cm long. Flowers are commonly reddish; there are many varieties with different flower colours (even white).
A. scherzerianum—Generally more adaptable as an indoor plant than *A. adraeanum*. Stems are generally very short, flowers commonly scarlet. Many hybrids produced from this species.

FAMILY Asteraceae
GENUS *Argyranthemum* syn. *Chrysanthemum*
COMMON NAME Marguerite daisy

Appearance: Roundish shrubs to 1m with fern-like, deeply lobed leaves. Daisies appear in profusion over the plant in season.
Establishment: Full sun and fertile soil. Can tolerate wind and dry air. Grows well in the wet tropics but flowering may be greatly reduced.
Culture: Fertilise in early spring and after flowering. Keep moisture up to the plant during flowering. Prune after flowering to remove spent flowers and to reshape plant. Prone to mildews if foliage is kept moist.
Cultivars
Many varieties are commonly grown in both temperate and hot climates, varying in both form and flower colour.

FAMILY Liliaceae
GENUS *Aspidistra*
COMMON NAME Cast iron plant

Appearance: Large, strappy, broad leaves arising from underground rhizome.
Establishment: Select a site in shade and protected from strong, dry winds.
Culture: Water and fertilise regularly to keep growth rate up. Prune only damaged leaves.If grown in the ground, control invasive habit in the tropics by removing a proportion of the plant each season. Attacked by grasshoppers, scale and mites. Propagate by division.
Cultivars
A. elatior—Commonly grown as a container plant in temperate climates, and occasionally as a spreading clump or border in the tropics and sub-tropics. The flower is obscure, but the foliage to around 50cm tall provides a

lush green mass. There is also a less vigorous yellow variegated form.

FAMILY Begoniaceae
GENUS *Begonia*

Appearance: Soft-tissued plants with highly colourful flowers.

Establishment: Select a protected but sunny site. Prefers a fertile, well drained soil.

Culture: Water and fertilise to maximise growth and flowering. Prune to maintain shape and spread. Prone to mildews and spider mites. Propagate by seed and cuttings.

Cultivars

Begonias have been bred extensively to produce a wide range of named varieties cultivated from temperate to tropical climates. They are generally classified into four groups:

Fibrous-rooted begonias. This is a large group, including plants that will grow across a wide variety of climates from mild temperate areas into the tropics. They include *B. sempervirens* (the common bedding begonia) and the taller tree begonias, which grow into small bushy clumps.

Rhizome types. These have thick root-like rhizomes that creep over the soil surface and sprout leaves periodically. They like warmth and humidity.

Rex begonias. These are varieties derived from the species *B. rex*, which is native to Assam. They typically have large, heavily textured, colourful foliage. They do best in a warm humid environment.

Tuberous-rooted begonias are more commonly grown in temperate climates.

FAMILY Apocynaceae
GENUS *Catharanthus*
COMMON NAME Periwinkle syn. Vinca or Ammocallis

Appearance: Semi-woody bushy plant with terminal flowers and green leaves, often with a white or light-coloured central vein.

Establishment: Select a site with full sun and fertile soils. Propagate by seed and cuttings.

Culture: Water and fertilise to maximise flowering. Prune to prevent weedy habit and to improve shape and maximise flowering. Problems include mealy bugs, caterpillars, grasshoppers.

Cultivars

C. roseus—This is perhaps the most commonly cultivated species, with reddish flowers to 70cm tall. Other varieties also exist, with different flower colours.

FAMILY Lamiaceae
GENUS *Coleus*

(Recent taxonomic revision of this genus has reclassified some *Coleus* into the genus *Solenostemon*).

Appearance: Small soft-tissued shrubs with highly attractive multi-coloured leaves.

Establishment: Full sun to part shade. Protect from strong winds.

Culture: Water regularly in dry or windy periods. Fertilise with slow-release pellets or liquid formulations. Tip prune to encourage bushiness and to remove flower heads. Main pest problems are from grasshoppers, caterpillars, leafhoppers.

Cultivars

C. blumei and other species have been developed and interbred producing many named varieties with different coloured foliage from red and purple to yellow and green. These may grow to around 50cm tall, and are used as bedding plants for temporary displays in temperate to tropical climates. Several other species are also widely grown in warmer climates, including *C. ambonicus* (grey-green foliage to 70cm tall) and *C. repens* syn. *C. carnosus* with grey-green creeping foliage.

FAMILY Gesneriaceae
GENUS *Episcia*
COMMON NAME Carpet plant

Appearance: Creeping herb, with hairy or even velvety foliage. Leaves oval to elliptical shape, tubular flowers often colourful.

Establishment: Prefers high humidity and temperature, shade, a fertile organic and moist soil. Does well in containers.

Culture: Water well when growing, never allow to dry out. Mulch. Prune after flowering.

Can suffer root rots in cooler conditions. Spray fungicide in wet, humid conditions.

Cultivars

Around 10 species, and many named varieties exist. The most commonly grown cultivars are varieties of *E. cupreata* (most of these have dark green coppery coloured foliage).

FAMILY Asteraceae
GENUS *Gerbera*

Appearance: Underground stem structure produces green, partially lobed leaves that lie horizontally, with star-like flowers in bright colours that face upward.

Establishment: Full sun and moist, fertile soil. Add plenty of organic matter to the soil.

Culture: Water in dry periods — do not allow the plant to suffer drought stress as this reduces flower production. Fertilise with high bloom boosting mixtures. Prone to mildews in high humidity situations. Propagate by division of crown or seed.

Cultivars

Gerberas have been widely bred for the cut flower industry. There are many named varieties including most of the colours in the rainbow.

FAMILY Acanthaceae
GENUS *Hypoestes*
COMMON NAME Polka dot plant

Appearance: Ground covers and small shrubs.

Establishment: Part shade preferred in a warm, sheltered site. Loves damp places. Can become a weed, self seeding or suckering, so often better suited to container growing.

Culture: Water in dry periods. Prune to control spreading habit. Propagation by seed or cuttings.

Cultivars

H. phyllostachya is the most commonly grown species, to 45cm tall. Soft-textured leaves covered with pink spots, pale lilac flower spikes.

FAMILY Balsaminaceae
GENUS *Impatiens*
COMMON NAME Busy Lizzy

Appearance: Soft-tissued plants with upright or trailing habit. Colourful leaves.

Establishment: Full sun to full shade, protected from the wind. Full shade will not produce prolific flowering.

Culture: Water in dry periods. Fertilise regularly. Prune to improve shape. Problems include grasshoppers, caterpillars, mites, mildews. Propagate by seeds or cuttings.

Cultivars

Impatiens have been bred extensively to produce six main types:

Singles are the easiest to grow. These have a single row or whorl of flower petals. As such, the flowers are not as spectacular as some types, but when planted as a mass they create a stunning effect from a distance. Singles are easy to grow from seed (usually the only type available as seed). The self-seeding capacity can mean they sometimes become a weed.

Semi-doubles have more complex flowers with more petals in each flower (but not as many as a true double).

Doubles have many petals in each flower, and when removed from the foliage, the flowers look similar to a miniature rose or African Violet flower. They are more susceptible to problems than the singles (eg. attack by mites, which can cause distortion of foliage).

New Guinea hybrids often have darker green foliage and more vibrant flower colours than other types. Leaves are longer and more pointed. They grow well in humid tropical areas, but are more temperamental elsewhere.

Pixies have masses of tiny flowers, and are a relatively hardy type.

Special types are other varieties which are less commonly grown. Some of these are quite different to other *Impatiens*, and may be even more spectacular.

FAMILY Amaranthaceae
GENUS *Iresine*
COMMON NAME Bloodleaf

Appearance: Soft-stemmed, spreading, rambling plant with colourful leaves.

Establishment: Part shade preferred with plenty of fertile, moist soil. Protect from strong winds. Propagate by cuttings.

Culture: Water in dry periods. Fertilise twice a

year. Prune to control rambling habit, or to create a hedge. Prone to grasshoppers, leafhoppers and caterpillars.

Cultivars

I. herbstii (Beefsteak plant) — Commonly grown in the sub-tropics and tropics. It can grow to 1.8m, but is often smaller.

FAMILY Crassulaceae
GENUS *Kalanchoe*
COMMON NAME Palm Beach bells

Appearance: Succulent foliage, low growing, often spreading herbs and shrubs, some with attractive foliage and colourful flowers.

Establishment: Easily grown in pots or in the ground. Prefer full sun or light shade. Water freely when growing. Be careful, though, as some have become a weed in warmer climates. Propagate by seed, stem or leaf cuttings.

Culture: Control spread with cultivation or spraying. Feed and water if in pots.

Cultivars

K. blossfeldiana—To 30cm. Brilliant coloured flowers, grown as a bedding plant. Many named varieties available.

K. longiflora—To 70cm tall, yellow flowers.

K. pinnata—From 30cm to 1.5m tall, reddish flowers.

K. tubiflora—To 80cm tall, reddish flowers.

FAMILY Nyctaginaceae
GENUS *Mirabilis*
COMMON NAME Four o'clocks

Appearance: Soft-stemmed, erect plant with terminal flowers that open at dawn and die by sunrise. Large underground tubers developed to support plant through dormant phase.

Establishment: Protect from exposure, if necessary with tree guards, and mulch to maintain moisture. Propagate by seeds, cuttings, tuber separation.

Culture: Water in dry periods, fertilise twice a year. Prune to control spreading habit, and to remove spent flower heads to remove seed bodies. Pests include grasshoppers.

Cultivars

The most commonly grown species is *M. jalapa* syn. *M. uniflora*, (Four o'clocks), to 80cm tall with scented flowers in a range of colours and variations.

FAMILY Berberidaceae
GENUS *Nandina*
COMMON NAME Sacred bamboo

Appearance: Bamboo-like plant with clumping/spreading habit and upright stems with leaves predominantly terminal. Flower clusters held at apex. Attractive orange-red berries.

Establishment: Will grow in full sun to part shade in an open area, tropics to temperate zone. Propagate by semi-hardwood cuttings.

Culture: Relatively hardy from the tropics to temperate climates, provided soil remains moist (excessive wet or dry can be a problem). Few pests. Can be pruned lightly for shape. Likes some mulching and feeding perhaps annually.

Cultivars

There is only one species, which can grow to 2.5m tall, but the most common cultivar is a low-growing form:

N. domestica nana (Dwarf sacred bamboo)—Tight-growing, deciduous plant to 60cm with pinnate leaves. Striking autumn colours before leaf fall.

FAMILY Piperaceae
GENUS *Peperomia*
COMMON NAME Radiator plant

Appearance: Spreading or clump-forming, low-growing plants with fleshy leaves.

Establishment: From the tropics and sub-tropics; needs high humidity and warm temperatures to do well. Some grown as indoor pot plants in cooler climates. Prefer filtered light to shade, and a fertile organic soil.

Culture: Water and fertilise regularly. Prune off dead leaves, or to control shape. Can suffer from sucking and chewing insects occasionally, and susceptible to root rots if waterlogged, or if roots remain moist when temperatures drop. Propagate easily by leaf cuttings.

Cultivars

Many species and named cultivars are grown as container plants, or in the tropics as shade plants in the ground.

FAMILY Urticaceae
GENUS *Pilea*
COMMON NAME Artillery plant

Appearance: Small annual or perennial herbs, with fleshy stems and often decorative oval-shaped leaves. Some species are trailing, forming a dense mat; others are small upright shrubs.

Establishment: Likes warm, humid conditions. Does well out of direct sunlight. Does well in pots. Prefers good drainage, even sandy soils.

Culture: Prune freely to control size and shape. Mulch to maintain soil moisture. Responds well to feeding and watering during dry periods. Propagates easily from cuttings.

Cultivars

More than 200 species exist, including many named varieties that are widely grown as indoor plants in temperate climates, and shade plants in hotter climates.

FAMILY Piperaceae
GENUS *Piper*
COMMON NAME Pepper

Appearance: Commonly spreading shrub, but sometimes other habits. Foliage has a pungent odour. Leaves are simple (undivided) and usually heart-shaped; flowers are on a central spike.

Establishment: Prefers high humidity, warm temperatures and a very rich but well drained soil. Propagate by cuttings.

Culture: Prune to control size. Mulch, feed and water freely as required.

Cultivars

Commercial black and white pepper come from the species *P. nigrum*, which grows as a vine in southern India and Sri Lanka. Several other species are grown as ornamental plants in the tropics and sub-tropics.

FAMILY Lamiaceae
GENUS *Plectranthus*
COMMON NAME Swedish ivy, Prostrate coleus

Appearance: Soft, often hairy foliage, leaves often toothed and sometimes scented; foliage colour variable, including species with purplish, greenish and bluish leaves. Often a spreading shrub, throwing roots out wherever succulent stems fall close to the ground. Most have blue-lilac to white flowers, reminiscent of coleus or salvia.

Establishment: Many are sensitive to extreme cold (though some grow well in temperate climates). *Plectranthus* are generally easy to grow in a moist, protected semi-shaded position in a mild to warm climate. Propagate readily by cuttings.

Culture: Sensitive to extreme dry. Cuttings can rot easily if they sit in water, but established plants are usually hardier. Prune freely to control spread and encourage bushy growth.

Cultivars

Many tropical species have been cultivated for a long time. Several Australian native species (mostly with scented bluish foliage) are being increasingly cultivated, showing great potential for the tropics and sub-tropics.

FAMILY Liliaceae
GENUS *Protasparagus* syn. *Asparagus*
COMMON NAME Asparagus fern

Appearance: Ornamental fern-like foliage with succulent root system.

Establishment: :Adapted to full sun and medium shade. Tolerant of various soil types but grows best in fertile, drained soils. Some species can naturalise and become a weed if uncontrolled.

Culture: Water in dry periods. Fertilise annually. Prune to control size and spread, and to remove 'fronds' that die back. Propagate by seed or division.

Cultivars

Several species of *Asparagus* are commonly seen as sprawling ground covers in tropical climates, including, *P. densiflorus sprengeri*, which has fine foliage and can reach 1.8m tall if left uncontrolled.

FAMILY Commelinaceae
GENUS *Rhoeo*
COMMON NAME Purple-leafed spider wort, Oyster plant

Appearance: Short rosette-shaped plant with striking foliage that is green on top and purple beneath.

Establishment: Though native to temperate

Asia, it is grown widely in the tropics and sub-tropics. Likes full sun to part shade, and a drained, moist organic soil.

Culture: Water and fertilise regularly, Prune only if essential. Once established it is very hardy, suffering neglect well. Propagate by division.

Cultivars

Only one species, *R. spathacea*, but there are a number of cultivars, including a variegated form.

FAMILY Scrophulariaceae
GENUS *Russelia*
COMMON NAME Cigarette plant, Coral plant

Appearance: Arching limbs produced from a central crown that fall to the ground due to weight. Deep green 'leafless' stems carry hundreds of small cigarette-like flowers.

Establishment: Prefers full sun to part shade, in exposed or protected site. Propagate by cutting and division of crown.

Culture: Tolerant of a range of soils and water conditions. Prune only if essential.

Cultivars

R. equisetiformis syn. *R. juncea*—This is the most widely grown species, to 1m tall with spreading, fine, weeping foliage and clusters of bright red flowers.

FAMILY Liliaceae
GENUS *Sansevieria*
COMMON NAME Mother-in-law's tongue, Snake plant

Appearance: Underground rhizome produces erect, stiff leaves. Foliage most commonly 20-90cm tall, often variegated in cultivated species.

Establishment: Does well in sun or shade, but can suffer sunburn in direct hot sun. Prefers a drained organic soil. Can become a weed in tropical and sub-tropical climates, one plant eventually spreading over very large areas.

Culture: Susceptible to root rots in cooler conditions (roots should be kept drier as weather gets cooler). Prune or thin out freely to control spread. Can suffer scale and some other insects, but generally relatively few pests.

Cultivars

Around 60 species. Some of the more commonly grown ones include:

S. cylindrica—To 1.2m tall, leaves with dark green stripes.

S. trifasciata—To 1m tall. Several named varieties with different leaf colourings and heights.

FAMILY Cactaceae
GENUS *Schlumbergera*
COMMON NAME Christmas cactus

Appearance: Epiphytic cactus. Foliage is flattened, jointed stems, more like a succulent than most other cacti. Flowers can be brilliant, and in a variety of colours.

Establishment: Grows well in containers or hanging baskets in light shade or full sun, in warm humid or dry climates. Will also grow in well drained dryish soil. Suits humid tropics to mild temperate areas.

Culture: Avoid overwatering in cooler weather. Responds to light feeding and watering in hot weather. Propagate by cuttings.

Cultivars

Many named varieties are grown, with both single and double flowers ranging from whites to pinks and reds.

FAMILY Selaginellaceae
GENUS *Selaginella*

Appearance: Primitive non-flowering plant, more like a moss than a flowering plant. Typically soft, tender fern-like foliage, with green or blue-green colouring.

Establishment: Prefers moist or wet organic soil in heavy or light shade. Often grown as a border or ground covering; can be grown as a hedge. Grows in most soil types.

Culture: Keep well watered and mulched in dry periods. Easy to propagate by division, layers or cuttings.

Cultivars

There are over 700 species, and many do very well in the humid tropics.

S. involvens—To 30cm tall, with dense foliage.

S. uncinata—To 90cm tall. Likes warm, semi-shaded positions.

S. willdenovi—Has a climbing habit and blue-green foliage.

FAMILY Araceae
GENUS *Spathiphyllum*
COMMON NAME Spathe flower, Spathe lily

Appearance: Clusters of leaves emerge from short, spreading rhizomes from a lush green clump. Flowers normally white, often fragrant.

Establishment: Likes moist organic soil in hot, humid conditions; some species will grow in waterlogged conditions. Likes medium to well lit conditions, but avoid direct sunlight. Excessive shade may reduce flowering. Will grow in temperate areas if roots are kept dry in cooler weather.

Culture: Can suffer from scale or mealy bug, otherwise generally trouble-free in tropical climates.

Cultivars

There are many named cultivars commonly grown, including *S. clevelandii* syn.. *S. wallisii* 'Clevelandii', to 30 cm or taller; and *S. friedrichsthalii*, to 1m tall.

9

Bulbs and Herbaceous Perennials

FAMILY Gesneriaceae
GENUS *Achimenes*
COMMON NAME Magic flowers

Appearance: Low-growing fleshy herbs with showy tubular flowers at tips of shoots.

Establishment: Prefers part shade and humidity. Good in hanging baskets. Propagate by division of rhizomes.

Culture: Keep moist at all times, fertilise regularly. Susceptible to thrip and red spider mite. Provide plenty of humidity leading up to flowering; reduce watering after flowering.

Cultivars

A. candida—reddish stems, serrated leaves, flowers purple-spotted throat with white and yellow petals.

A. hybrid—slender stems that creep horizontally then bend upwards. Many flowers over summer and autumn. Variety of flower colour forms available.

FAMILY Amaryllidaceae
GENUS *Agapanthus*

Appearance: Green strap-like leaves, heads of white or blue flowers on stalks above foliage. Old forms display flowers 1m above plant. New forms available as dwarfs.

Establishment: Add compost into soil to encourage vigorous growth. Does well in full sun to medium shade. Propagate by division.

Culture: Tolerates some drought but responds to water during dry periods. Fertilise twice annually, once in spring and once in autumn. Attracts snails and slugs; few serious problems. Keep tidy by cutting off old flower spikes.

Cultivars

A. africanus—Strap-like leaves 2cm wide, blue flowers in clusters up to 30.

A. orientalis—Strap-leaves 5cm wide, flowers in clusters up to 110, usually on 60cm long stalks.

A. 'Tinkerbell'—dwarf, variegated form.

A. africanus minor 'Peter Pan'—dwarf form with pale blue flowers on 50cm long stalks.

FAMILY Araceae
GENUS *Alocasia*
COMMON NAME Elephant ears, Ape plant

Appearance: Clump-forming with large heart-shaped leaves arising, often up to 2m high, from central growth rhizome.

Establishment: Add moisture-retentive compost or peat into the site. Provide protection from strong sun and wind in early years; will tolerate full sun provided ample moisture is present. Needs a frost-free area. Propagate by seeds, offshoots, division of rhizome.

Culture: Keep site moist, add compost and mulch regularly. Pests can include cluster caterpillar, thrips.

Cultivars

A. macrorrhiza (Ape, Biga)—Large arrowhead-shaped ornamental leaves borne on plant up to 4m. Although edible, it must be treated before eating.

A. macrorrhiza 'Variegata'—large or small irregular splotches of creamy-white or grey-green on the leaves.

A. macrorrhiza 'Rubra'—Pale violet leaf blades.

Many hybrid forms are also available.

FAMILY Zingiberaceae
GENUS *Alpinia*
COMMON NAME Gingers

Appearance: Fleshy pseudostems arising from underground rhizomes. Pseudostems display alternating leaf blades. Small flowers are borne in clusters of bracts either close at ground level or at the tips of pseudostems.

Establishment: Apply plenty of peat or compost to the site. Additional fertiliser beneficial. Provide protection from wind and full sun in early stages. Select a protected moist and humid site. In low humidity areas, select a dappled-light site and water the foliage. Avoid frosts. Propagate by division of rhizomes.

Culture: Keep moist to encourage vigorous growth and long flowering period. Responds well to mulching, Few pests (occasionally grasshoppers) and diseases.

Cultivars

A. purpurata (Red ginger)—To 2–5m tall, 30cm long red flower spike, best in the true tropics.

A. sanderae (Variegated ginger)—To 45cm tall, white and green leaves.

A. zerumbet (Shell ginger)—To 4m tall, white and yellow terminal, weeping flower clusters.

A. zerumbet 'Variegata'—Tough glossy yellow and green leaves.

FAMILY Amaryllidaceae
GENUS *Alstroemeria*
COMMON NAME Peruvian lily

Appearance: Thinly rhizomed plant with erect, soft, fleshy shoots displaying colourful flowers.

Establishment: In hot climates, rhizomes must be stored in the refrigerator for a month or so in winter, or brought in from cooler climates (ie. at least 4 weeks below 5°C is needed to produce flowers). Improve soil site with organic matter and fertiliser. Plant rhizomes in spring. Best in friable moist soil, full sun. Propagate by dividing rhizomes.

Culture: Water frequently and liquid fertilise every 4 weeks when in growth. Reduce water in winter. Harvest by pulling stems (not cutting), which stimulates development of more flowers.

Cultivars

Approximately 60 species, varying from miniatures to clumps 1.8m tall. Many medium to taller varieties can be grown successfully as cut flowers

FAMILY Amaryllidaceae
GENUS *Amaryllis*
COMMON NAME Hippeastrum, Belladonna

Appearance: Stocky, bulbous plant with strap-like foliage, sometimes deciduous. Large trumpet-shaped flowers (usually clusters of 3–5) borne on strong upright stalk, usually in late spring.

Establishment: Full sun or very light shade, good drainage. Tolerates wide range of climates except heavy frosts, prefers higher humidity. Plant bulbs when dormant, or potted specimens when in growth. Roots grow best at soil temperatures from 22–30°C. Propagate by division.

Culture: Tolerates poor and dry soil conditions once established, but will do better with some moisture and fertilisation. Responds to mulching (to keep soil warm). Generally pest-free.

Cultivars

There is only one species (*A. belladonna*), which is very similar to the genus *Hippeastrum* and often treated as being equivalent to *Hippeastrum*.

FAMILY Zingerberaceae
GENUS *Amomum*

Appearance: Scented rhizomes. Gives rise to linear to lance-shaped leaves to 30cm long and 7.5cm wide, and cone-like flower spikes. Some produce spicy, edible seeds.

Establishment: Grown as foliage plants in tropics and sub-tropics. Must have a fertile and moist soil, preferably with lots of humus. Propagate by division in spring.

Culture: Responds to watering and mulching in dry periods. Feed annually to increase production.

Cultivars

Over 100 species

A. Cardamomum syn. *Elettaria cardamomum* (Cardamom ginger)—Source of edible cardamom. Fragrant leaves. Shoots up to 3m. Dense cluster growth habit.

A. compactum (Round cardamom) to 1.8m tall.

FAMILY Araceae
GENUS *Amorphophallus*
COMMON NAME Devil's tongue, Snake palm
Appearance: Herbaceous plant with upright, lobed leaves commonly to 1m across, that die down in winter to a flattened tuber. Male flowers can have an unpleasant scent. If corms are strong, a flower is produced soon after planting and leaves emerge soon after that.
Establishment: Requires fertile, friable soil, part shade. Grows well in pots. Propagate by division or seeds.
Culture: Responds to mulching, regular feeding, and watering when in active growth. Dry corms over winter and store at 12°C or higher over winter.
Cultivars
Approx. 90 species, often grown as a 'curiosity'.

FAMILY Iridaceae
GENUS *Aristea* syn. *Witsenia*
COMMON NAME Blue brilliant
Appearance: Herbaceous, fibrous-rooted perennials. Leaves mainly from the base, sometimes up to 1m or more tall. Flower clusters usually blue, iris-like flowers, often in spikes.
Establishment: Prefers friable organic soil, likes good ventilation and plenty of light. Older plants lack vigour in their roots. Propagate by seed or division.
Culture: Transplanting best, taking divisions from young plants in winter.
Cultivars
Cultivated species include:
A capitata syn. *A. coerulea* — To over 1m tall, blue flowers, from South Africa.
A. corymbosa—To around 1m, purple flowers, from South Africa.
A. ecklonii— 50cm–1m tall, bright blue flowers.

FAMILY Iridaceae
GENUS *Babiana*
COMMON NAME Baboon flower
Appearance: Cormous, hairy green foliage. Flowers often fragrant, commonly reds, purples or whites.
Establishment: Prefers light sandy soils, full sun or light shade. Plant 5cm apart. Mainly from South Africa. Propagate by offsets.
Culture: Keep plants dry when and if growth dies down in cold weather, and replant in spring (avoid frost). If necessary, irrigate from time new season's growth starts until flowers decline.
Cultivars
Approx 61 species. Cultivated species include:
B. plicata—To 20cm, pale blue, purple or white flowers.
B. stricta—To 30cm tall, bluish flowers. Most commonly grown variety.
B. stricta var. *sulphurea*—To 30cm tall, yellow flowers. Commonly grown variety.
B. stricta var *rubrocyanea*—Normally around 15cm, scarlet to royal-blue flowers.
B. villosa—To 35cm tall, claret to purplish flowers.

FAMILY Amaryllidaceae
GENUS *Boophone*
COMMON NAME Buphone, Buphane
Appearance: Bulbous, perennial herb, strap-like leaves.
Establishment: Prefers full sun, does well in sub-tropics. Good in rockeries or dry slopes, will grow well in pots. Propagate by offsets.
Culture: Water freely when growing, but stop watering after leaves turn yellow.
Cultivars
B. ciliaris — To 30cm tall, purple flowers.
B. disticha — To 30cm tall, purple flowers.

FAMILY Amaryllidaceae
GENUS *Brunsvigia*
Appearance: Bulbous herb. Strap-shaped leaves emerge after the red or pink flowers.
Establishment: Prefers fertile, sandy soil, full sun and heat. Propagate by division.
Culture: Only water when growth starts; best kept dry during dormant period. Grows best in a very hot position, but needs to be kept cool during dormancy. For best results, grow in pots or lift bulbs at the end of the season, so they can be stored cool until the next year.
Cultivars
Approx. 20 species, cultivated species include:
B. gigantea—To 30cm, red flowers.

B. josephinae—To 50cm, scarlet flowers.
B. kirkii—To 50cm.

FAMILY Araceae
GENUS *Caladium*
Appearance: Tuberous roots, deciduous in cool months; ornamental arrow-shaped leaves up to 1m tall.
Establishment: Prefers protection from wind and drought, in filtered sun or shade. Do best in the humid tropics to sub-tropics, in the ground. Propagate by division.
Culture: Keep moist when in active growth. Hold back watering in dormant phase, keep well fertilised in growing season. Susceptible to mealy bugs and red spider mites.
Cultivars
C. bicolor—Has great variation in leaf patterns and colours. Small, insignificant aroid flowers borne close to the soil.

FAMILY Marantaceae
GENUS *Calathea*
COMMON NAME Zebra plant, Rattlesnake plant
Appearance: Tufts forming from colourful leaves that arise from ground level. Some species produce edible tubers.
Establishment: Prefers warm, humid climate, protection from wind and cold, temperatures rarely below 20°C. Light to heavy shade; well drained but moist organic soil. Propagate by division.
Culture: Responds well to mulch and compost; needs watering during dry periods. Few serious pests.
Cultivars
Approx 100 species. Cultivated species include *C. insignis* (to 1.7m), *C. lancifolia* (to 50cm), *C. louisae* (to 80cm), *C. makoyana* (to 60cm), *C. ornata* (1-2.5m), *C. zebrina* (40-80cm).

FAMILY Amaryllidaceae
GENUS *Calostemma*
Appearance: Bulbous, flowering Australian perennial.
Establishment: Prefers full sun or very light shade, in friable, freely draining soil. Grows well in pots. Propagate by offsets.
Culture: Responds well to mulching and irrigating in dry periods. Keep drier when dormant.
Cultivars
Cultivated species include:
C. album—To 30cm tall, white flowers.
C. luteum—To 30cm tall, yellow flowers.
C. purpureum—To 30cm tall, purplish flowers.

FAMILY Cannaceae
GENUS *Canna*
COMMON NAME Canna lily, Indian shot
Appearance: Spreading clump with large, commonly green or purplish leaves. Masses of colourful flowers on heads rising above foliage.
Establishment: Very adaptable, provided soil is reasonably moist when growing. Full or filtered sun is best, some drainage and fertilisation is appreciated. Does not tolerate heavy frosts. Can be propagated by seed if hard seed coat is broken with a knife or file before planting.
Culture: Clumps will spread fast and grow taller in a friable, fertile organic soil. Mulch and irrigate during dry periods. Cut back to ground level and divide after flowering.
Cultivars
Approx 50 species, and many cultivars, varying in height, flower colour and leaf colour. Widely grown from temperate to tropical climates.

FAMILY Amaryllidaceae
GENUS *Clivia*
COMMON NAME Kaffir lily
Appearance: Clumps forming bulb-like structures (on some species) at base of thick strap-like leaves, fleshy roots. Flower stalk rising above foliage bares a head of reddish or yellow flowers.
Establishment: Grow in pots or the ground. Does well in temperate or subtropical climates; may grow well but seldom flower in humid tropics. Prefers friable, moist soil. Loves shade. Propagate by seed or division.
Culture: Generally very hardy. Responds well to mulching and feeding. Water during dry periods.

Cultivars
Species grown include:
C. miniata (scarlet or yellow flowers, to 40cm).
C. nobilis (red or yellow flowers, to 40cm).

FAMILY Araceae
GENUS *Colocasia*
COMMON NAME Taro
Appearance: Heart-shaped leaves in a spreading clump.
Establishment: Prefers a protected, shaded site with high humidity, best in tropics and subtropics. Propagate by division of tubers.
Culture: Keep moist, but well drained. Add plenty of compost into soil. Fertilise in growing season. Can suffer from leafhoppers. Note: Tubers only edible after cooking.
Cultivars
C. esculenta—Edible tubers bearing large aroid leaves, to 2m or more tall.
C. sp. (Variegated Colocasia)—A smaller plant to 1.5m.

FAMILY Costaceae
GENUS *Costus*
Appearance: Fleshy, upward-growing pseudostems, producing leaves in radiating spiral habit. Flowers mostly small in showy bracts, produced terminally.
Establishment: Prefers highly organic, moist soil and partial shade in protected sub-tropical and tropical site. Propagate by division of rhizomes, sometimes by plantlets developed at end of pseudostems.
Culture: Keep moist. Susceptible to grasshoppers. Responds well to mulching, and annual applications of compost and rotted manure.
Cultivars
C. spiralis (Scarlet spiral flag)—To 1-3m; red-crimson bracts with red flowers.
C. spicatus (Indian head ginger)—To 2.5m; stiff, orange-red bracts with small yellow-orange flowers.
C. malortieanus (Stepladder plant)—To 1m; broad leaves to 30cm long, yellow flowers marked with red.
C. speciosus (Spiral ginger)—To 3m; tends to droop in graceful arches. Glossy leaves, silky undersides, red bracts with yellow flowers.
C. productus—To 1.5m; glossy leaves to 15cm long, stiff open orange bracts with orange flowers.
C. igneus (Fiery Costus)—To 1m; stout, strap leaves with large orange flowers.
C. sanguineus (Violet spiral flag)—Coppery-green stem with red petioles and velvety blue-green leaves, deep blood-red beneath.

FAMILY Amaryllidaceae
GENUS *Crinum*
COMMON NAME Spider lily
Appearance: Herbaceous plants with bulbous bases. Strap-like leaves radiate out around the base of the central flower spike which produces spider-like flowers in clusters.
Establishment: Prefers full sun to part shade; will tolerate sea wind exposure. Hardiness varies according to variety. Propagate by seed or root offshoots.
Culture: Keep well watered, but well drained; don't water during dormant periods. A rich, fertile soil preferred. Susceptible to cluster caterpillars and spider mites.
Cultivars
C. amabile (Sumatran giant lily)—To 1.6m tall and wide; green-red leaves 1–1.4m long. Pinkish flowers trimmed with red, and red filaments and anthers.
C. asiaticum (Giant lily)—To 2.4m tall and wide. Bright green leaves 2m long, pure white flower petals with red filaments and yellow anthers. Will tolerate dry conditions.
C. augustum (Queen Emma lily)—To 2.4m tall and wide; 2m long green leaves, wine-coloured petals with red filaments and yellow anthers. Good in hot, dry, sunny sites.

FAMILY Iridaceae
GENUS *Crocosmia*
COMMON NAME Montebretia, Monbretia
Appearance: Cormous herb. Similar to *Tritonia*.
Establishment: Prefers sandy soil, full sun, good border plant.
Culture: Generally as for *Gladiolus*; corms can be left in the ground for several seasons. Propagate by seed or offsets.

Cultivars
C. aurea 'Coppertip'—To 1m tall, yellow-orange flowers.
C. crocosmaeflora (*C. aurea* X *C. potsii*)—To 80cm tall, orange-red flowers.

FAMILY Zingiberaceae
GENUS *Cucurma*
COMMON NAME Turmeric, Curry
Appearance: Clump-forming tuberous roots. Flowers usually nestled in coloured bracts. Similar growth habit to gingers.
Establishment: Likes moist, well drained organic soils in well-lit position. Propagate by division.
Culture: Responds to mulch, and annual applications of compost and rotted manure. Grows well in Darwin, Cairns and south to Brisbane.
Cultivars
Cultivated species include:
C. albiflora, to 60cm; *C. longa* (Tumeric), to 70cm; *C. petiolata*, to 50cm; *C. roscoeana*, to 30cm, and *C. australis*, to 2m.

FAMILY Amaryllidaceae
GENUS *Cyrtanthus*
COMMON NAME Ifafa lily, Fire lily
Appearance: Bulbous plant. Fragrant red or white tubular flowers in clusters on top of flower stalks which rise a little above foliage.
Establishment: Likes friable, moist, well drained soil. Grows well in pots. Frost-sensitive, does best in frost-free subtropical areas. Prefers good light; avoid heavy shade. Mainly from South Africa. Propagate by fresh seed, or division.
Culture: Water freely when growing; reduce watering when growth slows. Responds to feeding annually.
Cultivars
Species cultivated include:
C. angustifolius, to 45cm, red flowers
C. mackenii, to 30cm, white flowers.

FAMILY Commelinaceae
GENUS *Dichorisandra*
COMMON NAME Blue ginger
Appearance: Fleshy stems with glossy, green leaves and clusters of blue or purplish flowers (upright or drooping).
Establishment: Adapted to a range of soils and sun factors. Protect from strong dry winds. Prefers medium to high humidity, prefers tropics to sub-tropics. Propagate by root division, stem cuttings.
Culture: Keep moist in warmer seasons, though soil must be well drained. Fertilise in growing season to encourage abundant flowers. Can suffer from thrips and mites. Prune spent flower heads and to control plant habit.
Cultivars
D. reginae—To 70cm tall, blue flowers with a white base.
D. thyrsiflora (Blue ginger)—To 2m in the tropics and partial shade. Only reaches about 1m in full sun. Vivid blue flowers produced in clusters at apex of stems, above glossy green leaves.

FAMILY Iridaceae
GENUS *Dierama*
COMMON NAME Wandflower
Appearance: Summer flowering corm, often tall, sometimes arching foliage. Flowers rarely erect—usually drooping.
Establishment: Plant in full or filtered sun, avoid cold. Good in sub-tropics. Prefers friable soil. Propagate by dividing bulbs.
Culture: Responds well to mulching. Lift corms over winter every few years.
Cultivars
Species grown include:
D. gracile—To 20cm, brownish flowers,
D. pendulum—To 1.8m, purplish to whitish flowers with brownish markings.
D. pulcherrimum— To 1m, red flowers.

FAMILY Amaryllidaceae
GENUS *Eucharis*
COMMON NAME Amazon lily, Eucharis lily
Appearance: Bulbous plant up to 40cm tall, with fleshy, deep green, glossy leaves.
Establishment: Prefers a shady site with protection from frosts and dry winds, with fertile, well drained soil. Propagate by separation.
Culture: Water well during growing season, provide a bit of a dry period during autumn.

Fertilise regularly when in active growth and flower, feed well once flowers start. Susceptible to minor spider mite damage and thrips.

Cultivars

E. grandiflora—Stunning white, daffodil-like flowers borne in spring or autumn above deep green foliage.

Family Liliaceae
Genus *Eucomis*
Common name Pineapple lily

Appearance: Bulbous plant with fleshy roots, Foliage similar to a pineapple. Flowers form in cylindrical, usually pale neutral-coloured clusters with sometimes a tinge of a brighter colour.

Establishment: Generally hardy in warm climates. Prefers well drained organic soil. Does well in containers, full or filtered sun. Propagate by division or seeds.

Culture: Responds well to mulch: water in dry weather while growing.

Cultivars

Cultivated species include:

E. bicolor—To 80cm, greenish flowers with purplish margins.

E. comosa syn. *E. punctata*Leaves spotted with purple underneath; flowers cream and green sometimes with a pink tinge.

Family Iridaceae
Genus *Gladiolus*

Appearance: Spring and summer flowering corm, cluster of strap-like leaves, one tall flower spike from each corm, containing many colourful flowers.

Establishment: Prefers moist, freely draining soil while growing, full sun (shade can result in reduced flowers), protection from strong winds. Stake if necessary. Grows well in subtropical south-east Queensland.

Culture: For best results reduce watering or lift dormant corms over winter. Ventilation is important in very humid areas (high humidity can result in disease attacking flowers). Aphis can be a problem, transmitting a serious virus. Insect problems are more likely in low humidity. Propagate by dividing corms.

Cultivars

Many hundreds of named cultivars are commonly grown by commercial growers and home gardeners throughout the sub-tropics and mild temperate climates. Flower colours are extremely variable.

Family Zingiberaceae
Genus *Globba*

Appearance: Delicate ginger-like plant with upright pseudostems, and terminal flowers that tend to weep downward. Tends to die off to dormancy during winter.

Establishment: Prefers a protected warm, humid site. Best intropical areas, although some success in sub-tropics as a garden plant.

Culture: Keep moist in growing season. Liquid fertilise during growing and flowering. Responds to mulch and annual applications of compost and rotted manure. Propagate by division of rhizomes.

Cultivars

G. atrosanginea—To 1m tall, graceful stems with erect red bracts and yellow corolla.

G. bulbifera—To 0.3m tall, broad leaves, flowers bold yellow bracts.

G. schomburgkii—To 0.5m tall. Intricate yellow flowers with red spots.

G. wintii—To 1m tall with delicate arching pink/purple bracts.

Family Liliaceae
Genus *Gloriosa*
Common name Climbing lily

Appearance: Tuberous rooted climbers, deciduous, spectacular and colourful lily flowers produced continually through warm weather.

Establishment: Easy to grow in most soils, provided moist. Prefers tropical or subtropical, humid climate; grows well in filtered light to medium shade. Frost-sensitive. Needs something to climb on (eg. a tree or trellis). Propagate by dividing tubers. Will also self-seed readily.

Culture: Responds well to feeding, mulching and watering if at all dry.

Cultivars

G. carsonii—To 2.5m tall, yellow and brown flowers.

G. rothschildiana—To 2m tall, red and yellow flowers.

G. superba (Glory Flower)—To 3m tall, yellow and red flowers.

G. virescens—To 1.5m tall, yellow and red flowers.

FAMILY Amaryllidaceae
GENUS *Habranthus* (sometimes included under *Zephranthes*)

Appearance: Bulbous plant similar to *Zephranthes* with heads of pink, yellow or red flowers.

Establishment: Prefers light, sandy loam, a sunny position. A good rockery or border plant. Avoid frost.

Culture: Responds well to water and feeding when growing. Lift and replant when clumps start to deteriorate, or to propagate by division.

Cultivars

H. andersonii var. *texanus*—To 10cm tall, clumping, grass-like habit with bulbous flower.

H. brachyandrus—To 30cm tall, with single erect flower atop stem.

H. robustus (H. tubispathus)—Bold orange and yellow, large flowers to 8cm long.

FAMILY Amaryllidaceae
GENUS *Haemanthus*
COMMON NAME Blood lily; Red cape tulip

Appearance: Small deciduous, bulbous plant with strap-like leaves, capable of growing up to 1m tall, with reddish flowers borne on sturdy stalks.

Establishment: Prefers sunny or partial shade. Adaptable, but best in fertile, well drained soils in tropical and subtropical areas. Propagate by root offshoots or seed.

Culture: Keep moist in growing season. Provide a dry rest period when dormant, fertilise when in growth. Can suffer from thrips.

Cultivars

H. coccineus—Normally has 2 large, broad leaves to 70cm long, followed by blood red flowers after leaves die back.

H. multiflorus (Blood lily)—Blood-red flowers in ball-like cluster produced above a leafless stem; leaves flat and medium green.

H. natalensis—To 30cm tall, green, purple and yellow flowers.

FAMILY Zingiberaceae
GENUS *Hedychium*
COMMON NAME Ginger lily

Appearance: Tough rhizome produces pseudostems with terminal flower heads. Will die down to dormant rhizome in cool winters.

Establishment: Needs protection from strong, drying winds. Will grow in protected temperate climates, but prefers humid tropics to subtropics. Propagate by division of rhizome.

Culture: Keep moist during growing stages; responds well to mulch and annual fertilising. For neat appearance, remove pseudostems that have flowered. Few pests or diseases.

Cultivars

H. coccineum (Scarlet ginger lily)—To 2m tall, flowers with red to yellow bracts, and pink filaments.

H. coronarium (Butterfly lily, Garland flower)—Leafy stout stems to 2m, beautiful white flower.

H. densiflorum—Stems to 0.5m tall, colourful and dense flower spike with clustered bracts.

H. flavum (Yellow ginger)—To 1.5m tall, fragrant long yellow/orange flowers.

H. gardnerium (Kahili ginger)—To 1.5m tall, with prominent yellow bracts containing long red filaments.

H. greeni—To 2m tall, dense leaves with chaotic red to orange flowers.

H. x *kewense*—To 2.5m tall, cane-like, large, strap leaves with clustered spike.

H. thyrsiforme (Ginger lily)—Broad leaves that have colourful drooping terminal bract.

FAMILY Heliconiaceae
GENUS *Heliconia*

Appearance: Similar to banana plants in habit, ranging in heights from 30cm to many metres. Terminal flowers develop to weep downward or grow erect throughout warmer seasons.

Establishment: Likes full sun in the tropics,

partial shade in the sub-tropics. Protect from strong drying winds. Propagate by division of rhizomes.

Culture: Water well in growing and flowering season. Responds well to organic fertilisers, compost and mulch. Suffers from some leaf-chewing insects (eg. grasshoppers).

Cultivars

Hundreds of cultivars now exist, some of the more common are:

H. bahai—Includes many forms and subspecies.

H. caribaea—Robust upright plant to 6m with thick rhizome. Very large inflorescence to 30cm in waxy-yellow, upright habit.

H. longa—To 4m tall, with bright red to orange swan-like bracts with magnificent gold sepals.

H. mariae—To 4m tall, large densely packed red bracts that hang heavily.

H. psittacorum (Parrot's beak heliconia)—Upright plant to 2m tall, with brightly colour bracts and variable, black-tipped orange flowers. Old foliage usually dies back in winter. Many colour cultivars available.

H. rostrata (Hanging parrot's beak)—To 3m tall in shade. Pendulous green and yellow-tipped red bracts up to 50cm long, small, inconspicuous, yellow flowers.

FAMILY Amaryllidaceae
GENUS *Hippeastrum*
COMMON NAME Amaryllis

Appearance: Bulbous plant with flattish green leaves produced after large bell-like flowers borne on a sturdy stalk. Very similar to *Amaryllis*.

Establishment: Treat the same as *Amaryllis*.

Culture: Treat the same as *Amaryllis*.

Cultivars

Many named varieties varying in flower characteristics (eg. colour, singles, doubles, etc) have been bred for cut flowers and collectors.

FAMILY Amaryllidaceae
GENUS *Hymenocallis*
COMMON NAME Spider lily, Ismene

Appearance: Strap-shaped leaves, summer flowering bulbs, flowers commonly fragrant white or yellow.

Establishment: Very adaptable and hardy plant in tropical or subtropical climates. Enjoys full sun and exposed sites; grows well in pots. Propagate by division of bulbs, or seed.

Culture: Susceptible to cluster caterpillars, lily caterpillar, thrips, spider mites.

Cultivars

Many species and named hybrids are grown, including:

H. calathina syn. *H. narcissiflora*—Foliage to 70cm long, flowers pale yellow or green to white. Flowers resemble a daffodil (in shape) with spidery protrusions.

H. littoralis—Has white petals, white based-webbed filaments that change to green with red-orange anthers; pure green strap-like but short leaves or variegated forms.

FAMILY Amaryllidaceae
GENUS *Hypoxis*
COMMON NAME Star grass

Appearance: Stemless herb. Corm-like, short rhizomes, narrow leaves, white or yellow star shaped flowers.

Establishment: Generally easy to grow, preferring a dryish soil. Propagate by division.

Culture: Keep bulbs relatively dry when dormant, gradually increase watering when growth begins. Responds well to feeding and mulch.

Cultivars

Cultivated species include:

H. hirsuta—To 30cm tall, bright yellow flowers.

H hygrometrica—To 15cm tall with spreading leaves to 30cm long, yellow flowers.

FAMILY Iridaceae
GENUS *Iris*
COMMON NAME Iris

Note: Most irises are better suited to temperate climates. Some, however, do grow well in warmer climates.

Appearance: Strap-like leaves, attractive colourful flowers, producing bulbs or rhizomes.

Establishment: Prefer moist to wet soil, full sun or very light shade. Grows well from mild temperate climates to the equator.

Culture: Keep moist in growing season, allow

a drier phase during dormancy. Fertilise well in spring and late summer. Propagate by division of rhizome or bulbs.

Cultivars

Species which come from warmer climates include:

I. helenae (from Egypt), *I. hexagona* (from southern USA), *I. imbricata* (from south-west Asia), *I. missouriensis* (central USA into northern Mexico), *I. planifolia* (Spain and northern Africa).

FAMILY Zingiberaceae
GENUS *Kaempferia*

Appearance: Herbaceous perennials with fragrant, colourful flowers. Mainly from tropical Africa and Asia.

Establishment: Prefers moist, friable, well drained, organic soil. Propagate by division in late winter.

Culture: Water freely while growing, otherwise keep fairly dry.

Cultivars

Species cultivated include:

K. gilbertii (variegated leaves, to 30cm).
K. kirkii (15cm).
K. rotunda (to 30cm).

FAMILY Liliaceae
GENUS *Kniphofia*
COMMON NAME Torch lily, Club lily, Tritomas

Appearance: Rhizome-forming, usually stemless, forming clumps of grass-like foliage. Clusters of reddish or yellow flowers occur on poker-like spikes.

Establishment: Hardy from mild temperate to tropical climates. Likes full sun. In very hot climates, mulch to keep roots cooler. Propagate by seed or division.

Culture: Responds to watering in extended dry periods. Responds to feeding and mulching, but will generally survive neglect.

Cultivars

Several species and different named cultivars are commonly grown in various parts of the world, the most common being *K. uvaria* (to 1.2m tall).

FAMILY Liliaceae
GENUS *Lilium*
COMMON NAME Lily

Appearance: Bulb producing an unbranched single stem with alternate or whorled leaves and either one or several flowers at the top.

Establishment: Some grow better than others in warm climates. Though not commonly thought of as a tropical plant, there are commercial nurseries in subtropical climates which specialise in *Liliums*. Some support is normally required for stems (ie. stakes or trellis). Most prefer lots of light and a moist, friable and fertile organic soil. Propagate by division or cuttings of bulb scales.

Culture: Responds well to feeding, watering, and mulching. Bulbs are best lifted annually, and stored in the refrigerator over winter (in the crisper—not too cold). Replant in spring.

Cultivars

Species suited to warmer climates include *L. longiflorum*, *L. candidum*, *L. lancifolium.*

FAMILY Amaryllidaceae
GENUS *Lycoris*
COMMON NAME Spider lily

Appearance: Bulbs with narrow leaves at the base. Dies back before nerine-like flower appears.

Establishment: Plant bulbs 70% of their depth. Water moderately until foliage appears, then increase watering. Plant in filtered sunlight (not dark!). Likes an organic, well drained soil. Propagate by dividing bulbs.

Culture: Mulch and feed annually. Lift bulbs every few years.

Cultivars

Several species are cultivated, including:

L. aurea—To 30cm, yellow flowers.
L. radiata– To 30cm tall, dull reddish flowers.

FAMILY Marantaceae
GENUS *Maranta*

Appearance: Similar to *Calathea* with very minor differences in flower, although *Maranta* generally has a lower growing habit.

Establishment: As for *Calathea.*

Culture: As for *Calathea.*

Cultivars

M. arundinaceae (Arrowroot)—To 1.8m tall, white flowers.

M. bicolor—To 30cm tall, olive-green foliage.

M. leuconeura—To 30cm tall, multicoloured leaves green, white and purple.

FAMILY Iridaceae
GENUS *Moraea* syn. *Dietes*
COMMON NAME Butterfly iris

Appearance: Cormous perennial herbs. Thin, sometimes sparse linear foliage; clusters of iris-like flowers.

Establishment: Very hardy in mild or warm climates; prefers light sandy soil, grows in organic soils if well drained. Avoid extreme cold. Propagate by dividing bulbs.

Culture: Requires little attention, but will respond well to feeding and watering occasionally.

Cultivars

M. bicolor syn. *Dietes bicolor*—To 1m tall, creamy flowers to around 8cm across, with dark brownish markings.

M. iridioides syn. *Dietes iridioides*—To 1m tall, white flowers with lavender markings.

FAMILY Zingiberaceae
GENUS *Nicolaia*
COMMON NAME Torch ginger

Appearance: Ginger-like plant, although much taller. Elaborate flowers similar to a torch in structure.

Establishment: Prefers partially shaded, protected humid site. Best in tropics, can flower in sub-tropics with extra care. Propagate by division of rhizome.

Culture: Keep moist, fertilise well in growing season; responds well to mulch, and annual applications of compost and rotted manure. Occasionally susceptible to grasshoppers. In Darwin it is said not to flower until the rhizome is about the width of a man's wrist.

Cultivars

N. elatior syn. *Phaeomeria magnifica* — To 5m tall, complex and vividly coloured clumped bracts atop slender leafy canes.

FAMILY Liliaceae
GENUS: *Ornithogalum*

Appearance: Mainly producing small to large bulbs. Evergreen clumps. Flowers occur clustered up a spike.

Establishment: Generally hardy, but prefers fertile soil and full sun. Propagate by division.

Culture: Harvest as lower flowers begin to open; others will open over a period of weeks in a vase. Responds well to mulch and well rotted manure.

Cultivars

Some species are hardy into temperate climates, others prefer the sub-tropics, including:

O. thyrsoides—To 30cm tall, white flowers.

O. arabicum—To 30cm tall, white flowers with black centre.

FAMILY Amaryllidaceae
GENUS *Phaedranassa*
COMMON NAME Queen lily, Gay lily

Appearance: Bulbous, with many fleshy leaves growing close to the ground; occasional clustered flower stalks.

Establishment: Protect from hot, dry winds, suits sub-tropics. Propagate by separation of bulbs, or seed.

Culture: Mulch, fertilise and water in dry weather. Protect from strong wind or cold.

Cultivars

P. carmiolii (Queen lily) — To 1m tall, long fleshy leaves with small clusters of crimson flowers.

P. chloracra — To 1m tall, fleshy lance-shaped leaves, pink flowers.

FAMILY Iridaceae
GENUS *Sparaxis*
COMMON NAME African harlequin flower, Wand flower

Appearance: Corms, leaves mainly basal and narrow; colourful flower heads.

Establishment: Prefers fertile, light sandy soil. Plant in autumn. Propagate by division.

Culture: Requires little attention once established, except thinning out and watering occasionally.

Cultivars

Species grown include:

S. grandiflora—To 60cm tall, yellow or purple flowers.
S. tricolor—To 45cm, red or pink and white flowers with yellow throat.

FAMILY Amaryllidaceae
GENUS *Sprekelia*
COMMON NAME Jacobean lily

Appearance: Bulbous herb, with large deep-red flowers (10–15cm) borne in autumn after dormant summer period.
Establishment: Prefers rich, well drained soil in full sun to partial shade. Propagate by separation of bulbs, or seed.
Culture: Keep well watered in growing season. Susceptible to thrips.
Cultivars
Only one species: *S. formosissima* (Jacobean lily).

FAMILY Taccaceae
GENUS *Tacca*

Appearance: Perennial herb with upright stem and large glossy broad leaves.
Establishment: Prefers a protected site in partial shade, in tropics or sub-tropics. Propagate by root division or seed.
Culture: Keep moist in open potting mix or soil. Keep soil fertile by adding compost; use liquid fertiliser for pot specimens. Susceptible to thrips. Remove spent flowers to encourage new flowers.
Cultivars
T. chantrieri (Bat Flower) — To 80cm tall, with large leaves 60cm long. Almost black flowers produced throughout warmer months, bearing long 'whiskers' and 'wings'.

FAMILY Amaryllidaceae
GENUS *Zephyranthes*
COMMON NAME Zephyr lily, Storm lily

Appearance: Narrow grass-like leaves forming a bulbous evergreen clump. Flowers can be white or yellow to reddish.
Establishment: Very hardy in tropical and subtropical climates. Likes full or filtered sunlight; adapts to most soils, but prefers some moisture. Propagates relatively fast and easily by division. Seed also germinates easily.
Culture: Requires little care; responds well to mulching and feeding occasionally.
Cultivars
Z. candida—Flowers are commonly white with a yellow central stamen.
Others commonly cultivated in different parts of the tropics and sub-tropics include:
Z. atamasco—To 30cm tall, white and purplish flowers
Z. grandiflora—To 30cm tall, rose-pink flowers
Z. tubiflora—Flowers deep orange.

FAMILY Zingiberaceae
GENUS *Zingiber*
COMMON NAME Ginger

Appearance: Typically clump-forming with large banana-like leaves. Flowers arise from ground level.
Establishment: Suits subtropical to tropical areas, moist fertile soils, full sun or well lit situation.
Culture: Responds well to mulch and annual applications of compost and rotted manure. Propagate by division of rhizome
Cultivars
Z. officinale (Common ginger)—Slender pseudo-stems to 1m with narrow leaves. Dense flower spike. Source of culinary ginger.
Z. zerumbet (Wild ginger)—Shoots to 60cm tall, slightly hairy beneath leaves. reddish bracts to 8cm appear in summer on 30cm long stalks. Bitter tasting rhizome.
Z. zerumbet 'Variegata'—Leaves variegated with cream; cone-like bracts are vivid red with small yellow flowers.

Appendix

CORRESPONDENCE COURSES

The author and contributors to this book, through the Australian Correspondence Schools, conduct a large variety of correspondence courses which are particularly appropriate to growing tropical plants, including:

PALMS AND CYCADS
ORCHID CULTURE
AFRICAN VIOLETS
INTERIOR PLANTSCAPING
AUSTRALIAN NATIVES
HOME LANDSCAPING

CERTIFICATES AND ADVANCED DIPLOMAS IN HORTICULTURE

Many courses are accredited through Australian accreditation authorities and recognised as satisfying requirements for admission as a member of the Australian Institute of Horticulture. Many options are offered to specialise (eg. in such things as Landscaping, Permaculture, Hydroponics, Crops or Turf).

The school also conducts many other courses such as 'Plant Propagation', 'Cut Flowers', 'Herbs', 'Cacti and Succulents', 'Ferns', and various other specialised courses.

Most shorter courses take about 2 hours per week for 12 months to complete.

VIDEOS

John Mason has also produced and presented a range of gardening videos, including 'Australian Native Plants', 'Identifying Plants', 'Designing a Garden', 'Identifying Herbs' and 'Plant Propagation'.These are available by mail order, or on loan through some libraries.

Further information, including course fees, can be obtained from:
Australian Correspondence Schools
264 Swansea Rd Lilydale, 3140
Ph: (03) 9736 1882 Fax: (03) 9736 4034
or
P.O. Box 2092 Nerang East, Qld. 4211.
Ph: (07) 5530 4855 Fax: (07) 5525 1728

websites http://www.qldnet.com.au/acs/hort
http://www.qldnet.com.au/acs
E mail acs@qldnet.com.au
acs@onthe.net.au

OTHER BOOKS

Other books that may be of interest to gardeners in the tropics and sub-tropics include:

Growing Ferns by John Mason, published by Kangaroo Press.

Growing Australian Natives by John Mason, published by Kangaroo Press.

Growing Herbs by John Mason, published by Kangaroo Press.

Growing Vegetables by John Mason and Rosemary Lawrence, published by Kangaroo Press.

Commercial Hydroponics by John Mason, published by Kangaroo Press.

Trees and Shrubs for Warm Places by John Mason, published by Hyland House.

Developing a Tropical Garden by John Mason, published by Hyland House.

Tropical and Warm Climate Gardening by John Mason, published by Bay Books.

Index

Numbers in **bold** type refer to colour plates (between pages 48 and 49).